Strategic and Tactical Considerations on the Fireground

STUDY GUIDE
Fourth Edition

By Deputy Chief James P. Smith, ret.
Philadelphia Fire Department

ISBN: 978-1-63491-957-9

Publisher James P. Smith
Editor James P. Smith
Cover and Book Design: Booklocker
Front Cover Photo: Greg Masi
Photograph of Author: Gerald Hilton
Key words: Firefighting, Strategy and Tactics, Command and Control, Building Construction, Building Collapse, Incident Management System, Decision Making, Firefighter Safety, Special Occupancies, Technical Operations.

Printed on acid-free paper.

Library of Congress Cataloguing in Publication Data
Smith, James P. (James Patrick), 1946-

Strategic and Tactical Considerations on the Fireground Study Guide, Fourth Edition, James P. Smith

1. Command and control of fires. 2. Fire Extinction. I. Smith, James P. Smith

ABOUT THE AUTHOR

James P. Smith is a retired Deputy Fire Chief with the Philadelphia, PA, Fire Department. He started his career in 1966 and retired in 2007 and held every civil service rank within the department. He currently serves with the Atlantic County All-Hazard Incident Management Team as their Safety Officer. He is an adjunct instructor at the National Fire Academy and the Emergency Management Institute in Emmitsburg, MD. He has developed programs on building construction, building collapse, firefighter safety, incident command, high-rise firefighting operations, church fires, and strategy and tactics, and presents seminars nationwide. He is published in various national publications and is contributing editor for *Firehouse* magazine, fire studies column. He is the author of the book Strategic and Tactical Considerations on the Fireground, fourth edition for which this study guide is written. He can be contacted at JPSmithPFD@ aol.com

DEDICATION

This book is dedicated to the fine firefighters that I have worked with in my career especially those who have lost their lives in the line of duty.

I have received advice and guidance from many individuals that have assisted me on and off of the fireground. This is especially true of my friend and aide for over 20 years Charlie Armstrong who was really the brains behind the scenes, Deputy Chief Bill Shouldis and my son Jim Smith, Chief of the Ocean City New Jersey Fire Department who have always been there to assist me in any endeavor and provided great advice in writing my textbook.

And to my grandchildren Ashley, Tyler, Justin, Brendan and Sarah who brighten my life by their presence.

DISCLAIMER

This publication is designed to provide information that might be useful in developing firefighting strategies and tactics. It is intended only as an informational reference volume, and the reader is expressly cautioned to use all safety precautions, and to take appropriate steps to avoid hazards when engaging in the activities described within.

Neither the author nor the publisher makes any representations or warranties of any kind, with respect to the materials set forth in this publication, express or implied, including without limitation any warranties of fitness for a particular purpose or merchantability. Nor shall the author or the publisher be liable for any special, consequential, or exemplary damages resulting, in whole or in part, directly or indirectly, from the reader's use of, or reliance upon, this material or subsequent revisions of this material.

TABLE OF CONTENTS

This study guide includes 900 questions from the book *Strategic and Tactical Considerations on the Fireground, Fourth Edition*. Guides 1, 2, 3, and 4 cover Chapters 1 through 5. Guides 5, 6, 7, and 8 cover Chapters 6 through 11. Guides 4 and 8 are meant to be a midterm guide for the covered chapters. Guide 9 is a final review of the entire book. There is a key at the end of each guide listing the correct answer and a reference page(s) in the text to review the data.

INTRODUCTION

This study guide is meant as an accompaniment to the book *Strategic and Tactical Considerations on the Fireground, Fourth Edition* written by retired Deputy Chief James P. Smith of the Philadelphia, PA, Fire Department and published by Pearson. It is not meant to be an all-inclusive text or to answer all-encompassing questions; it is meant to reinforce the text after it is read. In many cases the questions are narrow in design and emphasize specific points made within the text.

 I strongly recommend that the text be read first, and then this study guide be used to reinforce it.

Naturally there are many methods of studying. An excellent way is not to use the study guide as a test; instead, without reading each question, mark the correct answer as noted in the key at the end of each set of questions. Then study the question and the correct answer(s). This permits one to study only correct information. The premise is that if you test your knowledge by answering the questions as if taking a test, you may find yourself defending incorrect answers, which then could become part of your base of knowledge and lead you astray when confronted with an actual test.

In any case, use this study guide in the way that you feel will assist you in your firefighting endeavors, and above all else "Be Safe."

Study Guide 1 Chapters 1 through 5

1.   **The stage of fire when the oxygen or fuel starts to diminish is called:**
   a.   fully developed stage.
   b.   growth stage.
   c.   decay stage.
   d.   underdeveloped stage.
   e.   none of the above.

2.   **The action required when a backdraft situation is recognized is to**
   a.   shout backdraft loudly and run away from the building.
   b.   announce over the radio that a backdraft is about to occur.
   c.   provide adequate ventilation above the fire.
   d.   only a and b.
   e.   all of the above.

3.   1)   **Preincident planning is a method of gathering facts about a building, or a process within a building.**
   2)   **Preincident planning lets a fire department evaluate conditions and situations in its area of responsibility prior to an emergency.**
   a.   Both statements are true.
   b.   Both statements are false.
   c.   Only statement number one is true.
   d.   Only statement number two is true.

4.   1)   **Historical data enables a fire department to select the most critical properties or specific problems in our community that should be preplanned.**
   2)   **Preplanning data should assist a fire department in analyzing potential problems and developing a plan of action based upon what may occur.**
   a.   Both statements are true.
   b.   Both statements are false.
   c.   Only statement number one is true.
   d.   Only statement number two is true.

5.   1)   **Responders should review and update preplan information during site visits.**
   2)   **The scheduling of multiple dates for site visitations so all fire department members can visit special preplanned sites is not necessary since the data should be incorporated into company drills to permit a constant refresher for all members.**
   a.   Both statements are true.
   b.   Both statements are false.
   c.   Only statement number one is true.
   d.   Only statement number two is true.

6. **1) Wire-glass windows may crack from radiant heat.**
    2) Smoke-proof doors and smoke-removal systems can assist in minimizing evacuation problems.
    a. Both statements are true.
    b. Both statements are false.
    c. Only statement number one is true.
    d. Only statement number two is true.

7. **1) Resource utilization includes securing the services of outside agencies, such as the police, Red Cross, private security companies, public works, public health, utilities, or Federal, State, or local agencies.**
    2) Outside agencies should not be used for evacuation purposes at incidents attended by the fire department.
    a. Both statements are true.
    b. Both statements are false.
    c. Only statement number one is true.
    d. Only statement number two is true.

8. **Implementation of a preplan during a simulated exercise can be used to adjust the preplan. It is beneficial to know:**
    a. What problems the fire department encountered.
    b. If the community was involved in the exercise, were there any problems?
    c. Did the plant or facility find any discrepancies in the plan?
    d. Only a and b.
    e. All of the above.

9. **According to the National Fire Academy's fire flow formula, for a fire involving 25 percent of the first floor of a 3-story building that is 30-feet by 40-feet with exterior exposures on two sides the fire flow would be**
    a. 100 gallons per minute.
    b. 200 gallons per minute.
    c. 400 gallons per minute.
    d. 500 gallons per minute.
    e. None of the above.

10. **Accordingly to the National Fire Academy's fire flow formula, if an exterior exposure becomes involved in fire what amount of fire flow should be added to the original fire flow?**
    a. 10-percent of the original fire flow.
    b. 15-percent of the original fire flow.
    c. 25-percent of the original fire flow.
    d. 50-percent of the original fire flow.
    e. None of the above.

11. 1) If the National Fire Academy's fire flow requirements for water supply exceed the fire flow capability of available resources, a defensive mode of operation usually is required.
    2) Situations will occur where fire is attacking lightweight structural components and, though there is a sufficient water supply and resources at the scene, the conditions will be too dangerous for an offensive attack.
    a. Both statements are true.
    b. Both statements are false.
    c. Only statement number one is true.
    d. Only statement number two is true.

12. 1) Company officers can be successful if they praise their firefighters' good behavior publicly and criticize their mistakes privately.
    2) A company officer, when reviewing a misdeed with a firefighter, must ensure that the discussion focuses on the mistake that was made and not become a discussion of personalities.
    a. Both statements are true.
    b. Both statements are false.
    c. Only statement number one is true.
    d. Only statement number two is true.

13. 1) Safe procedures will occur on an incident scene if the company officer allows the firefighters to perform their assignments without interference from the company officer.
    2) The implementation of safety in practice evolutions will carry over to the emergency scene.
    a. Both statements are true.
    b. Both statements are false.
    c. Only statement number one is true.
    d. Only statement number two is true.

14. 1) Delegation permits subordinates to assume responsibility and to make decisions. It permits a supervisor to assess the skills of subordinates, and can lead to suggestions on how they can improve.
    2) Delegation is a necessary training process whereby company officers can learn the duties and responsibility of the chief officer.
    a. Both statements are true.
    b. Both statements are false.
    c. Only statement number one is true.
    d. Only statement number two is true.

15.   **1)**   **The ability to command an incident scene takes preparation and development on the part of the Incident Commander.**

          2)   **Command is a demanding position that often needs decisions to be made by a committee.**

          a.   Both statements are true.
          b.   Both statements are false.
          c.   Only statement number one is true.
          d.   Only statement number two is true.

16.   **1)**   **Leadership starts with the ability to possess self-discipline and one who can recognize potential incident problems.**

          2)   **When indecisive orders are issued, they leave doubt in the minds of those on the receiving end of those orders and can lead subordinates to question their validity.**

          a.   Both statements are true.
          b.   Both statements are false.
          c.   Only statement number one is true.
          d.   Only statement number two is true.

17.   **1)**   **Firefighters who realize they are a prime consideration of the Incident Commander in his or her decision making process will often give that extra effort to ensure success of their assignments.**

          2)   **The lessons learned from an incident in another jurisdiction are just as helpful in enabling us to prepare for a similar occurrence in our area.**

          a.   Both statements are true.
          b.   Both statements are false.
          c.   Only statement number one is true.
          d.   Only statement number two is true.

18.   **ICS allows emergency responders:**

          a.   An organizational structure adaptable to any emergency or incident to which response agencies would be expected to respond.
          b.   A system applicable and acceptable to users throughout the country.
          c.   Readily adaptable to new technology.
          d.   Ability to expand in a logical manner from an initial fire attack situation into a major incident.
          e.   All of the above.

19.   **ICS allows emergency responders:**

          a.   Basic common elements in organization, terminology, and procedures.
          b.   Implementation with the least possible disruption to existing systems.
          c.   Effectiveness in fulfilling all management requirements costs.
          d.   All of the above.
          e.   None of the above.

20. How many interactive components does ICS have?
a. 5.
b. 6.
c. 7.
d. 8.
e. 9.

21. **The ICS interactive components provide the basis for an effective ICS concept of operation. These components include:**
a. Common terminology.
b. Modular organization.
c. Integrated communications.
d. Unified Command structure.
e. All of the above.

22. **The ICS interactive components provide the basis for an effective ICS concept of operation. These components include:**
a. Consolidated action plans.
b. Manageable span of control.
c. Designated incident facilities.
d. Comprehensive resource management.
e. All of the above.

23. **ICS requires common terminology which includes:**
a. Major organizational functions.
b. Units that are predesignated and titled.
c. Terminology is standard and consistent.
d. Each incident should be named.
e. All of the above.

24. **Common terminology in ICS requires that common names are established and used for**
a. all personnel.
b. equipment.
c. resources conducting tactical operations.
d. for all facilities in and around the incident area.
e. all of the above.

25. 1) **When units are designated a function they will no longer use their standard call letters. They will adopt their new designation for all communications.**
2) **If Engine 1's officer is assigned as Division 1 Supervisor he or she will use Engine 1 as the call sign for all radio communications and not Division 1.**
a. Both statements are true.
b. Both statements are false.
c. Only statement number one is true.
d. Only statement number two is true.

26. 1) **The ICS organizational structure develops in a modular fashion from the top down at any incident.**
     2) **The functional areas, which are implemented as the need develops, are Command, Operations, Logistics, Planning and Finance/Administration.**
     a. Both statements are true.
     b. Both statements are false.
     c. Only statement number one is true.
     d. Only statement number two is true.

27. **The command function within ICS may be conducted in two general ways:**
     1) **Single command may be applied when there is no overlap of jurisdictional boundaries.**
     2) **Unified Command may be applied when the incident is within one jurisdictional boundary, but more than one agency shares management responsibility.**
     a. Both statements are true.
     b. Both statements are false.
     c. Only statement number one is true.
     d. Only statement number two is true.

28. **Every incident needs an incident action plan (IAP). Written IAP's usually are required when:**
     a. The incident is long duration or involve multiple operational periods.
     b. When multiple jurisdictions are involved in the response.
     c. It is required by agency policy.
     d. All of the above.
     e. None of the above.

29. **Span of control refers to the number of personnel reporting to any given individual. Optimal span of control in the ICS is?**
     a. Three.
     b. Four.
     c. Five.
     d. Six.
     e. Seven.

30. **Designated incident facilities include:**
     a. Command post.
     b. Incident base.
     c. Staging area.
     d. None of the above.
     e. All of the above.

31. **The Incident Command System is a combination of:**
    a. Personnel.
    b. Facilities.
    c. Equipment.
    d. Communications.
    e. All of the above.

32. **The Command Staff consists of:**
    a. Operations Officer.
    b. Safety Officer.
    c. Liaison Officer.
    d. All of the above.
    e. Only b and c above.

33. **An incident scene where there is a response of multiple agencies demands a point of contact person for these agencies. This function is best performed by the:**
    a. Safety Officer.
    b. Incident Commander.
    c. Liaison Officer.
    d. Information Officer.
    e. Finance Officer.

34. **A command post should be located**
    a. at a vantage point from which to view the incident.
    b. with a view of the front and the most critical of the two sides of the structure.
    c. with a view of the direction toward which a fire may spread.
    d. only a and b.
    e. all of the above.

35. **Dispatch can use the status report to**
    a. prepare for move-ups of companies.
    b. notify mutual-aid companies of the situation.
    c. pass along information to senior officers who may have response duties.
    d. only a and c.
    e. all of the above.

36. **A final status report should be given**
    a. when the chief decides to give the report.
    b. at the time a fire is placed under control.
    c. when the fire is completely extinguished.
    d. after overhaul is completed.
    e. during the salvage operation.

37. **A method must be in place to clear a radio band if an important message has to be given. Which of the following methods can accomplish this?**
  - a. Emergency traffic.
  - b. Mayday.
  - c. Priority.
  - d. Only a and b.
  - e. All of the above.

38. **In regards to a formal written incident action plan:**
  - a. Forms 201, 202, 203, 204, 205, 206 and 208 are all part of the plan.
  - b. Forms 201, 202, 203, 204, 205, 206, 208, 215 and 215A are all part of the plan.
  - c. Forms 202, 203, 204, 205, 206 and 208 are all part of the plan.
  - d. Forms 201, 215 and 215A are all part of the plan.
  - e. None of the above.

39. **The command sequence is based on how many levels?**
  - a. 2.
  - b. 3.
  - c. 4.
  - d. 5.
  - e. 6.

40. **Incident stabilization includes**
  - a. confining the fire to as small an area possible.
  - b. stabilizing a patient on a medical response.
  - c. stopping a leak at a hazardous materials incident .
  - d. only a and c.
  - e. all of the above.

41. **Size-up**
  - a. gathers information for the development of strategic goals.
  - b. is a mental process weighing all factors of the incident against the available resources.
  - c. can be looked at as solving a problem.
  - d. only b and c.
  - e. all of the above.

42. **A 360-degree walk-around**
  - a. can be accomplished by actually walking around the incident scene.
  - b. can be accomplished by driving around the scene in an apparatus or chief's vehicle.
  - c. is not necessary if not responding on a multi-unit response.
  - d. only a and b.
  - e. all of the above.

43. **Water supply may be supplemented by**
   a. having the Water Company increase the pressure on the hydrant systems.
   b. using a private hydrant system.
   c. having secondary hydrant systems.
   d. only b and c.
   e. all of the above.

44. **The most important factor of size-up is**
   a. water supply.
   b. sufficient number of firefighters.
   c. life safety.
   d. only b and c.
   e. all of the above.

45. 1) **Narrow aisles, high-piled stock and heavy smoke conditions assist firefighters in finding a fire.**
    2) **High ceilings can cause fires to go undetected for a longer period of time and allow fire extension.**
   a. Both statements are true.
   b. Both statements are false.
   c. Only statement number one is true.
   d. Only statement number two is true.

46. **An Incident Commander can request additional units for the following reasons.**
   a. To accomplish a specific assignment.
   b. To relieve units already operating at the scene.
   c. To remain in staging for anticipated problems.
   d. Only b and c.
   e. All of the above.

47. **When considering the size-up factor "weather"**
   a. reduced hydrant pressure may occur due to illegally opened hydrants.
   b. heat and high humidity can drain the strength of firefighters.
   c. extreme heat will require more frequent relief for firefighters.
   d. only b and c.
   e. all of the above.

48. **The size-up factor "height" should consider**
   a. any structure more than one story in height.
   b. the floors above the fire will pose a threat due to possibility of vertical spread of fire.
   c. attached or adjacent structures of equal or greater height as immediate exposures.
   d. only b and c.
   e. all of the above.

49.   In order to know how to fight a fire, the contents of the fire building must be determined. Signs on the exterior of the building can be helpful. Highly combustible stock produces a high rate of fire spread, and certain manufacturing processes create situations allowing flash fires or rapid spread. The previous statement would apply to which size-up factor?
   a.   Auxiliary appliances.
   b.   Exposures.
   c.   Weather.
   d.   Time.
   e.   Occupancy.

50.   **Time can affect our response due to**
   a.   rush hour traffic in and around cities.
   b.   seasonal shopping in downtown shopping districts and malls.
   c.   seasons of the year may have increased fire loading.
   d.   only a and b.
   e.   all of the above.

51.   **In regard to strategy:**
   a.   Problems identified through size-up can be solved by implementing the necessary strategies.
   b.   Many firefighters group strategy, tactics, and tasks together
   c.   Tactics achieve the strategies.
   d.   Only a and c.
   e.   All of the above.

52.   **In regard to strategy:**
   a.   It should be viewed as overall "goals" and "what" you want to accomplish.
   b.   It accomplishes the tasks portion of the command sequence
   c.   There is no specific definition.
   d.   Only a and b.
   e.   All of the above.

53.   **Exposure protection considers the potential for fire to involve**
   a.   internal exposures.
   b.   external exposures.
   c.   the immediate fire area.
   d.   only a and b.
   e.   all of the above.

54.   **Overhauling ensures that**
   a.   all fire has been extinguished.
   b.   areas where fire could still be burning are checked.
   c.   smoldering contents are removed to the exterior.
   d.   only b and c.
   e.   all of the above.

55.   **Constant reevaluation of an incident is necessary to ensure**

a. that the strategies, tactics and tasks are accomplishing the needed goals.

b. that the Incident Commander keeps busy until the fire is extinguished.

c. that legal ramifications will be met, and that the fire department will not be liable should any errors occur at an incident.

d. only a and c.

e. all of the above.

56.   **Crew Resource Management (CRM) was initially developed by:**

a. United States Marine Corps.

b. United States Army.

c. National Fire Academy.

d. Emergency Management Institute.

e. Airline industry.

57.   **In regard to hose-line placement in occupied buildings**

a. the first hose-line should be placed between the fire and the occupants.

b. the second hose-line can be used to back up the first hose-line.

c. the third hose-line can protect any secondary exits.

d. only a and c.

e. all of the above.

58.   **Probably no other component is as important in achieving success at an incident scene as the proper placement of the hose-lines. The Incident Commander must realize that**

a. our ability to control and extinguish a fire often is dependent solely on this factor.

b. the size of the fire will dictate the number and size of hose-lines required.

c. for an interior attack to be successful, units must be able to advance the hose-line to the seat of the fire.

d. only a and c.

e. all of the above.

59.   **In regard to hose-line usage:**

a. Ventilation allows the hose-line crew to advance into the fire area.

b. If ventilation has not occurred, the opening of the hose-line into a closed room or fire area will make it untenable to fight the fire.

c. When advancing a hose-line and fire is encountered, a solid stream should be played on the ceiling and quickly rotated around the room.

d. Only a and b.

e. All of the above.

60. **In regard to exposures**
 a. the ability to protect exposures will be in direct proportion to the distance between the exposure and the fire buildings.
 b. frame structures create the greatest challenge due to the large amount of combustible material.
 c. where adjoining buildings are encountered, the possibility of extension is greatly reduced.
 d. only a and b.
 e. all of the above.

61. **A hose stream operating at an exterior fire operation that is not hitting the fire**
 a. is ineffective.
 b. steals water from other appliances.
 c. causes unnecessary damage and can further weaken the structure.
 d. only a and c.
 e. all of the above.

62. **If a fire is beyond the ability of hand-lines to control, master streams can be used to knock down the fire in a "blitz" attack. After knockdown, the building can be checked for structural stability and, if sound, an interior attack then can commence. In this particular instance the mode of attack employed is referred to as**
 a. offensive.
 b. defensive.
 c. offensive/defensive.
 d. defensive/offensive.
 e. nonintervention.

63. **In determining strategy the Incident Commander must**
 a. take a proactive approach considering the potential for immediate fire spread.
 b. decide what can reasonably be accomplished.
 c. know what resources are needed.
 d. only a and c.
 e. all of the above.

64. **1)** **It is seldom necessary to stretch a hose-line larger than 1½-inch or 1¾-inch into dwelling units.**
 2) **The 1½-inch or 1¾-inch hose-line can be ineffective or less effective when fighting fires in commercial properties containing a heavy fire load.**
 a. Both statements are true.
 b. Both statements are false.
 c. Only statement number one is true.
 d. Only statement number two is true.

65. **The selection of the number and size of hose-lines must consider**
    a. the type of fire.
    b. the size of the fire.
    c. the mode of attack.
    d. only a and b.
    e. all of the above.

66. **The most effective method of fire control and life safety in any structure is:**
    a. A fully staffed and properly trained fire department.
    b. A water system with adequate sized water mains and a sufficient number of hydrants.
    c. A properly designed and installed sprinkler system.
    d. Aggressive and well trained firefighters.
    e. Fire departments that have specialized equipment and advanced training facilities.

67. 1) **A one-directional search of a room should be made by keeping close to the walls and reaching out as the room is encircled.**
    2) **Thermal imaging is of little help in search and rescue operations.**
    a. Both statements are true.
    b. Both statements are false.
    c. Only statement number one is true.
    d. Only statement number two is true.

68. **In regard to search and rescue, the rescuers should use**
    a. full protective gear.
    b. a hand tool, light, and portable radio.
    c. a protective hose-line is required of all teams.
    d. only a and b.
    e. all of the above.

69. **In regard to laddering**
    a. the front of the fire building should be reserved for the truck company.
    b. the first-due engine should either stop before reaching the front of the fire building or pull past it to permit access for the truck.
    c. the front of the fire building places the truck in a position for ready access of the portable ladders on the truck.
    d. only a and b.
    e. all of the above.

70. 1) **A ladder placed to gain access to a roof should extend at least three rungs above the roof.**
    2) **A ladder placed to play a stream of water from a hose-line into a window should have the head of the ladder resting on the wall above the opening.**
    a. Both statements are true.
    b. Both statements are false.
    c. Only statement number one is true.
    d. Only statement number two is true.

71. 1) **When the base of the ladder is set too close to the building, the top of the ladder will tend to pull away from the building as the climber nears the head or top of the ladder.**
    2) **When the base of the ladder is set too far from the building, the base will tend to creep or walk away from the building.**
    a. Both statements are true.
    b. Both statements are false.
    c. Only statement number one is true.
    d. Only statement number two is true.

72. **Many fire departments paint the tips of their ladders with fluorescent paint to**
    a. allow each truck company to identify its ladders.
    b. allow firefighters to locate the ladder easily when operating on a roof.
    c. to prevent deterioration of the tips of the ladders.
    d. only a and b.
    e. all of the above.

73. **In regard to ventilation:**
    a. It is a 'must' of structural firefighting.
    b. Addressing ventilation as an afterthought leads to poor fireground operations.
    c. If no civilians are endangered, ventilation should be done after a charged hose-line is laid.
    d. Only a and b.
    e. All of the above.

74. **When performing roof ventilation**
    a. skylights should be removed intact.
    b. falling glass from skylights can injure firefighters operating below.
    c. if unable to completely remove a skylight, it should be broken and pulled back onto the roof.
    d. only a and b.
    e. all of the above.

75.   **Negative ventilation is accomplished by**
a.   placing fans to push the smoke or toxic fumes from a building.
b.   using fog nozzles to push smoke or toxic fumes from a building.
c.   placing fans to pull the smoke or toxic fumes from a building.
d.   only a and b above.
e.   all of the above.

76.   **Fire burning through a roof that is too weak to support firefighters should**
a.   be saturated with water from ladder pipes and tower ladders to protect surrounding buildings.
b.   be allowed to burn through the roof to create a ventilation opening.
c.   be cooled with hose-lines from adjoining roofs.
d.   only a and c.
e.   all of the above.

77.   **1)   Overhaul is "the checking of a fire scene to determine that no fire remains." A close examination should be made to ensure that every location where hidden fire could still be burning is thoroughly searched.**
2)   A rekindled fire often is attributed to poor overhaul practices.
a.   Both statements are true.
b.   Both statements are false.
c.   Only statement number one is true.
d.   Only statement number two is true.

78.   **In regard to water damage encountered during firefighting and overhauling**
a.   the best way to minimize water damage is strict control of all hose-lines.
b.   when operating a hose-line the firefighter should open the nozzle when fire is encountered and close it after the fire has been knocked down.
c.   water damage should not be a concern of firefighters.
d.   only a and b.
e.   all of the above.

79.   **In regard to firefighter safety during overhauling:**
1)   Portable lighting should be used when overhauling building interiors during nighttime operations, though it is not necessary during daytime operations.
2)   It should be assumed that, since the smoke has lifted at a fire scene, it is safe to remove self-contained breathing apparatus.
a.   Both statements are true.
b.   Both statements are false.
c.   Only statement number one is true.
d.   Only statement number two is true.

80.   **In regard to concrete:**
- a.   It can spall under some conditions.
- b.   Cracks can weaken a concrete wall.
- c.   It will withstand any type of fire condition.
- d.   Only a and b.
- e.   All of the above.

81.   **When operating on an insulated bar joist roof, it is sometimes difficult to determine the location of a fire below since**
- a.   the usual visible signs found on a non-insulated roof may not be present.
- b.   it is a built up roof and it is further from the fire.
- c.   the bar joist are spaced further apart than 24-inches on center.
- d.   only a and c.
- e.   all of the above.

82.   **Ordinary constructed buildings contain walls constructed of**
- a.   concrete.
- b.   stone.
- c.   brick.
- d.   all of the above.
- e.   none of the above.

83.   **In ordinary constructed buildings the bearing walls normally will be**
- a.   the shortest walls in length.
- b.   the longest walls in length.
- c.   the front and rear walls.
- d.   the side walls.
- e.   none of the above.

84.   **Void spaces in ordinary constructed buildings are**
- a.   commonly found.
- b.   rarely found.
- c.   only found in balloon construction.
- d.   only a and c.
- e.   none of the above.

85.   **Heavy timber construction**
- a.   provides an excellent degree of fire resistance.
- b.   provides a minimal amount of fire resistance.
- c.   affords firefighters time for an aggressive attack on a fire.
- d.   only a and c.
- e.   none of the above.

86. **In regard to heavy timber construction**
- a. the exterior walls are of masonry and can be up to eight stories in height.
- b. larger structures contain fire walls.
- c. the floors are built to carry heavy loads.
- d. only b and c.
- e. all of the above.

87. **The heavy timber building**
- a. as built, is not prone to collapse.
- b. will withstand attack by fire and give firefighters time to control and extinguish a fire.
- c. may have been modified or be in a deteriorated condition.
- d. only a and b.
- e. all of the above.

88. **When fighting fires in heavy timber buildings**
- a. a fire above the first floor will be fought initially from the stairways.
- b. if the fire threatens the floor above, a hose-line should be immediately stretched to that location.
- c. there is no need to stretch hose-lines to back up those already in place.
- d. only a and b.
- e. all of the above.

89. **The characteristics of log frame construction are**
- a. logs are used in log cabin homes and small commercial buildings.
- b. the logs may be finished on the interior or covered with paneling or drywall.
- c. under heavy fire conditions the walls usually withstand the effects of the fire.
- d. only a and b.
- e. all of the above.

90. **Frame row dwellings as a rule**
- a. use the adjoining sidewalls as bearing walls.
- b. have more support under fire conditions than a free-standing building.
- c. will magnify the exposure problem due to the interconnection of buildings.
- d. only b and c.
- e. all of the above.

91. **The exterior walls of frame buildings**
- a. can include materials that will supply fuel to a fire.
- b. can include noncombustible finishes.
- c. can include stucco, stone facing, or brick veneer.
- d. only b and c.
- e. all of the above.

92.   **The parallel chord truss is used in**
- a. lightweight truss only.
- b. heavy timber truss only.
- c. lightweight and heavy timber truss.
- d. bowstring truss.
- e. none of the above

93.   **Which part of a structure is the most important factor in determining whether a building will fail under fire conditions?**
- a. Bearing walls.
- b. Floors.
- c. Roof.
- d. Nonbearing walls.
- e. None of the above.

94.   **Timber truss roofs that contain sloping hip rafters**
- a. have four bearing walls.
- b. use the side walls as the only bearing walls.
- c. can have a violent collapse of all four walls.
- d. only a and c.
- e. all of the above.

95.   **When lightweight building components are used in a building**
- a. they are dependent on proper installation and bracing.
- b. trusses are built with set bearing points.
- c. trusses fail more readily than conventional construction when under attack by fire.
- d. only a and c.
- e. all of the above.

96.   **Collapse of wooden "I" beams occurs due to weakening of the beam by**
- a. improper installation and alterations.
- b. openings in the web that are too large.
- c. openings in the web that are too close together.
- d. only a and b.
- e. all of the above.

97.   **A problem for responding firefighters is that lightweight building components are especially dangerous during two distinct time periods:**
- a. Under construction.
- b. Under demolition.
- c. During transportation to the construction site.
- d. Only a and b.
- e. All of the above.

98. **Interconnection of void spaces within buildings containing parallel chord truss assemblies**
    a. contains a sufficient amount of air to sustain a fire.
    b. is not a problem for firefighters.
    c. can combine horizontal truss voids with vertical voids.
    d. only a and c.
    e. all of the above.

99. **In buildings containing lightweight building components a void space fire**
    a. may not be recognizable to the firefighters operating above.
    b. may damage flooring where firefighters can easily fall through it.
    c. if suspected in any concealed space, that area should be opened quickly.
    d. only a and b.
    e. all of the above.

100. **There are a number of considerations for firefighters when confronted with a fire in a building that may use lightweight building components as structural members. They should**
    a. read a building for indicators that would denote the presence of a truss.
    b. realize draft-stopping may add fuel to a fire..
    c. recognize that triangular truss roof spaces are often used for storage.
    d. only a and c.
    e. all of the above.

ANSWER KEY Study Guide 1 for Chapters 1 through 5

Question	Answer	Page Reference	Question	Answer	Page Reference
1	C	6	26	A	39
2	C	9	27	A	39
3	A	12	28	D	39
4	A	13	29	C	40
5	C	14	30	E	40
6	A	14	31	E	42
7	C	15	32	E	59
8	E	16	33	C	60-61
9	B	17–19	34	E	75
10	E	18	35	E	76
11	A	19	36	B	77
12	A	20	37	E	79
13	D	22	38	C	85
14	A	23	39	D	91
15	C	25	40	E	92
16	A	25	41	E	93
17	A	26	42	D	93
18	E	38	43	E	99–100
19	D	38	44	C	100
20	D	38	45	D	101
21	E	38	46	E	101
22	E	38	47	E	104
23	E	38	48	E	104
24	E	39	49	E	105
25	C	39	50	E	105

ANSWER KEY Study Guide 1 for Chapters 1 through 5

Question	Answer	Page Reference	Question	Answer	Page Reference
51	E	106	76	B	161
52	A	106	77	A	162
53	D	107	78	D	164
54	E	108	79	B	167
55	A	109	80	D	173
56	E	112	81	A	176
57	E	119–120	82	D	178
58	E	121	83	B	180
59	E	122	84	A	182
60	D	123	85	D	183
61	E	124	86	E	183
62	D	125	87	E	184
63	E	133 – 134	88	D	185
64	A	135	89	E	187
65	E	135	90	E	188
66	C	138	91	E	189
67	C	142	92	C	190
68	D	144	93	C	191
69	E	146	94	D	194
70	A	147 – 148	95	E	195
71	A	148	96	E	199
72	B	149	97	D	201
73	E	152	98	D	201
74	E	153	99	E	202
75	C	157	100	E	204

Study Guide 2 Chapters 1 through 5

1. **Fires involving flammable liquids, combustible liquids, petroleum greases, tars, oils, solvents, lacquers, alcohols, and flammable gases are classified as:**
   a. Class A fires.
   b. Class B fires.
   c. Class C fires.
   d. Class D fires.
   e. Class E fires.

2. **1) A fire officer cannot use timeframes derived from timed evolutions or performance standards in assigning tactical operations when commanding a fire scene.**
   2) Fire departments should regularly schedule cross training of members normally assigned to an engine on the operations of a main ladder or tower ladder.
   a. Both statements are true.
   b. Both statements are false.
   c. Only statement number one is true.
   d. Only statement number two is true.

3. **Fire departments that use preplanning find**
   a. that it can mean the difference between success and failure.
   b. that it has little value should a fire occur within a building that has been preplanned.
   c. that though preplanning has value it doesn't permit anticipating potential problems.
   d. all of the above statements are true.
   e. none of the above statements are true.

4. **Fire protection systems within a building could include**
   a. standpipes.
   b. sprinklers.
   c. CO_2 systems.
   d. dry chemical systems.
   e. all of the above.

5. **When doing a preplan, potential situations should be analyzed thoroughly to determine**
   a. how the building's occupants and those in threatened exposures can be protected.
   b. what evacuation plans have been formulated.
   c. if occupants can be protected by other means.
   d. all of the above.
   e. only a and b.

6.   **There are certain facilities where, if an emergency occurs, it can have a direct impact on the immediate surrounding community. These facilities could include**

    a.   chemical plants.
    b.   refineries.
    c.   large water purification plants.
    d.   only a and b.
    e.   all of the above.

7.   **1)**   **When working with preplans, the addition of contingency plans for foreseeable problems and their incorporation into simulated exercises should be encouraged.**

    2)   **A preplan should be reviewed only when the plant personnel bring changes to the attention of the fire department.**

    a.   Both statements are true.
    b.   Both statements are false.
    c.   Only statement number one is true.
    d.   Only statement number two is true.

8.   **Accordingly to the National Fire Academy's fire flow formula; if other floors in a building are not yet involved, but are threatened by possible extension of fire, they should be considered interior exposures. What percent of the actual fire flow should be calculated for each exposed floor?**

    a.   10 percent.
    b.   15 percent.
    c.   25 percent.
    d.   50 percent.
    e.   None of the above.

9.   **In using the National Fire Academy's quick-calculation method to determine required fire flows, it is important to remember that**

    a.   the answers provided by this formula are approximations of the amount of water needed to control the fire.
    b.   the formula is geared to an offensive attack.
    c.   the accuracy of the formula decreases with fires involving over 50 percent of a structure.
    d.   the formula is an estimate of both the area of the building and the amount of fire involvement within the building.
    e.   all of the above.

10. 1) The company officer is the direct link for firefighters between middle and executive management and he or she must maintain a critical balance between them.
     2) Company officers must accomplish the goals of the department, while looking out for the well-being of their firefighters.
     a. Both statements are true.
     b. Both statements are false.
     c. Only statement number one is true.
     d. Only statement number two is true.

11. 1) Training is the backbone of every good organization. How well we practice dictates how well we will perform at an emergency. The more we train the better we become.
     2) Through the use of standard operating guidelines, we have the ability to practice the performance of routine tasks that then can be applied at the incident scene.
     a. Both statements are true.
     b. Both statements are false.
     c. Only statement number one is true.
     d. Only statement number two is true.

12. 1) The experience learned as a company officer will be the foundation that chief officers can build upon.
     2) A chief must assume responsibility for management and leadership in the fire department.
     a. Both statements are true.
     b. Both statements are false.
     c. Only statement number one is true.
     d. Only statement number two is true.

13. 1) At an incident the chief, after surveying the scene by doing a 360-degree walk-around, should enter the fire building and direct operations.
     2) The chief officer must rely on the company officer to be his or her eyes and ears in areas that cannot be seen by the chief.
     a) Both statements are true.
     b) Both statements are false.
     c) Only statement number one is true.
     d) Only statement number two is true.

14. **1)** **A leader who exhibits self-confidence will gain the confidence of his or her subordinates. This characteristic can be referred to as "command leadership."**

2) **The need for command presence is magnified at emergency scenes. Time cannot be wasted if people are in need of rescue. Immediate action is required.**

a. Both statements are true.
b. Both statements are false.
c. Only statement number one is true.
d. Only statement number two is true.

15. **When responding to an incident, an emergency vehicle should attempt to**

a. view the roadway as far ahead as we can see for possible problems
b. change traffic lanes to assist in your response
c. select a traffic lane and try to fully utilize it if possible
d. consider a parallel roadway or alternate route if gridlocked
e. all of the above

16. **During the operational period an incident action plan should cover:**

a. Strategies.
b. Tactics.
c. Support activities.
d. All of the above.
e. None of the above.

17. **Factors that impact upon span-of-control ratios include:**

a. Training/Experience level of subordinates.
b. Complexity of the incident.
c. Type or timeframe of the incident.
d. All of the above.
e. Only a and b.

18. **1)** **By dividing the incident into manageable segments, the Incident Commander is able to reduce the number of individuals directly reporting to him or her and is able to properly manage the incident.**

2) **Command officers must anticipate span-of-control problems and prepare for them--especially during rapid buildup of incident organization.**

a. Both statements are true.
b. Both statements are false.
c. Only statement number one is true.
d. Only statement number two is true.

19.   **Designated incident facilities can include:**
   - a.   Command Post.
   - b.   An Incident Base.
   - c.   A Staging Area.
   - d.   All of the above.
   - e.   Only a and b.

20.   1)   **The Command Post is the location from which all incident operations are directed.**
       2)   **Multiple command posts can be established per incident.**
   - a.   Both statements are true.
   - b.   Both statements are false.
   - c.   Only statement number one is true.
   - d.   Only statement number two is true.

21.   **Comprehensive resource management may be accomplished using:**
   - a.   Single resources.
   - b.   Task Forces.
   - c.   Strike Teams.
   - d.   All of the above.
   - e.   None of the above.

22.   1)   **The ICS has five major functional areas: Command, Operations, Planning, Logistics, and Finance/Administration.**
       2)   **Use of the ICS improves safety by providing proper supervision, accountability, coordinated efforts, and improved communications.**
   - a.   Both statements are true.
   - b.   Both statements are false.
   - c.   Only statement number one is true.
   - d.   Only statement number two is true.

23.   **Which Homeland Security Presidential Directive (HSPD) directs the Secretary of Homeland Security to develop and administer a National Incident Management System (NIMS)?**
   - a.   HSPD-1.
   - b.   HSPD-2.
   - c.   HSPD-3.
   - d.   HSPD-4.
   - e.   HSPD-5.

24.   **The National Incident Management System has how many components?**
   - a.   3.
   - b.   4.
   - c.   5.
   - d.   6.
   - e.   7.

25. 1) **Multiagency Coordination Systems may be required on large or wide scale emergencies that require higher-level resource management or information management.**
    2) **A Multiagency Coordination System is a combination of resources that are integrated into a common framework for coordination and supporting domestic incident management activities.**
    a. Both statements are true.
    b. Both statements are false.
    c. Only statement number one is true.
    d. Only statement number two is true.

26. **The primary functions of Multiagency Coordination Systems are to:**
    a. Support incident management policies and priorities.
    b. Facilitate logistics support and resource tracking.
    c. Make resource allocation decisions based on incident management priorities.
    d. Coordinate incident-related information.
    e. All of the above.

27. **Under NIMS the position of Intelligence/Investigations may be organized in which of the following ways?**
    a. Officer within the Command Staff.
    b. Unit within the Planning Section.
    c. Branch within the Operations Section.
    d. Separate General Staff Section.
    e. All of the above.

28. **Command responsibilities include:**
    a. Determine incident objectives.
    b. Determine tactics when there is an Operations Chief.
    c. Develop Incident Action Plan.
    d. Only a and c.
    e. All of the above.

29. **The definition of unity of command is**
    a. everyone has bosses.
    b. rank is not recognized on an incident scene.
    c. numerous people may report to one person.
    d. no one reports to more than one person.
    e. none of the above.

30. 1) **Cooperating and assisting agency representatives are assigned to the Liaison Officer.**
    2) **The agency representatives who interact with the Liaison Officer need to have decision making authority within their respective organization.**
    a. Both statements are true.
    b. Both statements are false.
    c. Only statement number one is true.
    d. Only statement number two is true.

31. **The use of staging at an incident scene**
    a. assists in the control of units.
    b. is a location from which units can be quickly deployed.
    c. is used to address the strategy and tactics of the Incident Commander.
    d. only a and b.
    e. all of the above.

32. **Once Command is established**
    a. it can then be forgotten.
    b. further implementation or usage is not necessary.
    c. its continuity must be maintained.
    d. only a and b.
    e. all of the above.

33. **The types of status reports are**
    a. initial, delayed, and ongoing.
    b. initial, ongoing, and final.
    c. delayed, initial, and ongoing.
    d. immediate, delayed, and final.
    e. immediate, ongoing, and final.

34. 1) **Communications is the giving and receiving of information and is the backbone of any emergency response organization.**
    2) **The Incident Commander must control communications and request regular updates from divisions and groups.**
    a. Both statements are true.
    b. Both statements are false.
    c. Only statement number one is true.
    d. Only statement number two is true.

35. 1) **The use of numbered codes is the preferred method of radio transmissions since it reduces radio traffic.**
    2) **It must be assumed that an order given via radio that has not been acknowledged has been received.**
    a. Both statements are true.
    b. Both statements are false.
    c. Only statement number one is true.
    d. Only statement number two is true.

36.   **The length of time for an "operational period" is determined by:**
- a.   Incident Commander.
- b.   Operations Section Chief.
- c.   Planning Section Chief.
- d.   Logistics Section Chief.
- e.   All of the above.

37.   **The command sequence**
- a.   can be used to assist in decision making.
- b.   is a standardized and sequential thought process.
- c.   works perfectly at every incident.
- d.   only a and b.
- e.   all of the above.

38.   **Property conservation means**
- a.   minimizing property damage.
- b.   quick extinguishment of a fire.
- c.   good salvage methods.
- d.   only a and c.
- e.   all of the above.

39.   **En route size-up will be based on**
- a.   information given to dispatch by those reporting the incident.
- b.   personal knowledge of the structure by the responders.
- c.   information on any preplans that may be in place.
- d.   only a and c above.
- e.   all of the above.

40.   **The initial company officer's size-up must determine**
- a.   the volume and intensity of the fire.
- b.   the initial strategies and tactics to be deployed.
- c.   the resources needed for overhaul.
- d.   only a and b.
- e.   all of the above.

41.   **When implementing a water tender operation consideration must be given to**
- a.   the amount of time involved.
- b.   the water available.
- c.   the terrain.
- d.   the travel time.
- e.   all of the above.

42.   **When considering life safety**
- a.   the most endangered will be those in the immediate vicinity of the fire area and those directly above the fire.
- b.   ventilation and the placement of hose-lines must take into account fire confinement and protection of trapped occupants.
- c.   never take occupants down fire department ladders.
- d.   only a and b.
- e.   all of the above.

43.   **When resources are limited, the Incident Commander**
- a. should handle each situation exactly the same as when ample resources are available.
- b. must adjust his or her priorities according to what can actually be accomplished.
- c. should always become a working firefighter since Command will not be a necessary function.
- d. only a and b.
- e. all of the above.

44.   **When a fire attacks structural components**
- a. the potential for collapse must be considered.
- b. the type of building components will vary and so will the collapse potential.
- c. it is important for firefighters to know the various building components.
- d. only a and c.
- e. all of the above.

45.   **When considering the size-up factor "weather"**
- a. the speed and direction of the wind can determine how a fire will travel.
- b. high winds make it difficult to control a fire.
- c. high winds can break down fire streams.
- d. only a and b.
- e. all of the above.

46.   **Many serious fires in office or commercial buildings in terms of loss of life and financial loss seem to be related to**
- a. delayed discovery.
- b. delayed notification of the fire department.
- c. employees trying to fight the fire.
- d. only a and b.
- e. all of the above.

47.   **In regard to size-up**
- a. it is an ongoing process.
- b. it assists in the handling of an incident scene.
- c. it is the basis for developing strategy and tactics.
- d. only a and b.
- e. all of the above.

48.   **Incident priorities are used**
- a. when the Incident Commander deems it necessary.
- b. only if a problem develops and analysis is needed.
- c. as a basis for decision making.
- d. only a and c.
- e. all of the above.

49.   **When considering exposures**
- a.   the protection of exposures is vital in containment and control efforts.
- b.   it must be determined what is exposed in adjacent areas.
- c.   it is helpful to know what is on the floors above.
- d.   only b and c.
- e.   all of the above.

50.   **Overhauling ensures**
- a.   complete extinguishment.
- b.   that utilities are shut down if needed.
- c.   that fire protection systems are restored if possible.
- d.   that a determination of the cause of the fire is initiated.
- e.   all of the above.

51.   **Problems facing the initial Incident Commander who is trying to achieve his or her strategy, tactics, and tasks are**
- a.   limited resources.
- b.   minimal information.
- c.   time constraints.
- d.   only a and b.
- e.   all of the above.

52.   **In addition to decisions on the mode of attack, the first-due company officer on arrival must decide**
- a.   what size hose-line will be needed.
- b.   where to place the first line.
- c.   where to place the second and third lines.
- d.   only a and c.
- e.   all of the above.

53.   **In regard to hose-line placement at fires in an unoccupied building:**
- a.   The first hose-line should attempt to cut off the fire spread.
- b.   The second hose-line can back up the first hose-line or go to the floor above.
- c.   The third hose-line then can be placed to back up either hose-line, or to contain horizontal or vertical fire spread.
- d.   Only a and b.
- e.   All of the above.

54.   **In regard to hose-line usage**
- a.   water played onto the ceiling will come down and knock down the fire in the room quickly and efficiently.
- b.   when the fire has been darkened down, the shutoff should be closed.
- c.   a single stream with an adjustable nozzle can knock down the fire first and then be switched to fog for hydraulic ventilation.
- d.   only a and c.
- e.   all of the above.

55. **The first hose-line for exposure protection in a defensive mode of attack**

    a.   should be placed to protect the most endangered property.

    b.   if life is threatened in the exposures, the exposure most critically endangered will be protected first.

    c.   if the first hose-line can't adequately protect the primary exposure, the second hose-line should back it up.

    d.   only b and c.

    e.   all of the above.

56. **Apparatus placement for a defensive attack must consider**

    a.   current collapse zones.

    b.   potential collapse zones.

    c.   anticipated incident needs in positioning engine and truck companies.

    d.   only b and c.

    e.   all of the above.

57. **The correct mode of attack to be deployed at an incident will result from the Incident Commander's size-up. As higher ranking officers arrive on the scene their assessment of the incident must consider whether to**

    a.   continue the strategies and tactics that are already in place.

    b.   modify the strategies and tactics in place.

    c.   change the mode of operation.

    d.   only b and c.

    e.   all of the above.

58. **When predicting fire conditions at an incident scene and considering "confinement" one must take into account**

    a.   the contents of the fire building.

    b.   the type of building construction.

    c.   the presence of fire protection features within a building.

    d.   only a and c.

    e.   all of the above.

59. **In regard to large diameter hose-line**

    a.   it allows large volumes of water to be moved from the source to the fire.

    b.   it can be used for relay operations.

    c.   it is of little value to fire department operations.

    d.   only a and b.

    e.   all of the above.

60. **Hose-line advancement requires**

    a.   sufficient ventilation in front of the nozzle.

    b.   a hose-line of sufficient size to control the fire.

    c.   four or more firefighters in the fire area.

    d.   only a and b.

    e.   all of the above.

61.   **A master stream is capable of water flows in excess of**
    a.   200 gallons per minute.
    b.   250 gallons per minute.
    c.   300 gallons per minute.
    d.   none of the above.
    e.   there is no set amount of gallons for a master stream.

62.   **In regard to search and rescue, the rescuer should**
    a.   use a right-hand or left-hand search pattern.
    b.   listen for sounds of victims.
    c.   check under and behind furniture.
    d.   only a and b.
    e.   all of the above.

63.   **In regard to search and rescue, the rescuers should**
    a.   recognize alternate escape routes from all interior areas.
    b.   memorize all outside features of the building.
    c.   identify different roof levels that would allow for easy exit from a window.
    d.   only a and c.
    e.   all of the above.

64.   **1)**   **The presence of overhead electrical wiring unseen by firefighters operating or climbing a main ladder should not be a concern since the ladder is grounded and it will not electrically energize the ladder should a firefighter come into contact with the wire.**
    2)   **Trees can interfere with raising ladders. A main ladder can sometimes be maneuvered through openings in the tree branches to reach windows or roofs.**
    a.   Both statements are true.
    b.   Both statements are false.
    c.   Only statement number one is true.
    d.   Only statement number two is true.

65.   **1)**   **When placing a ladder to ventilate a window, it should be placed on the leeward side of the opening, slightly above the top of the window.**
    2)   **A ladder placed to gain access to a roof should extend at least five rungs above the roof.**
    a.   Both statements are true.
    b.   Both statements are false.
    c.   Only statement number one is true.
    d.   Only statement number two is true.

66. **When placing a portable ladder**
    a. securing it ensures that the ladder will not be affected by high winds.
    b. the ladder should take advantage of cracks in concrete.
    c. setting a ladder in soft earth is beneficial since it will be secured by the dirt.
    d. only a and b.
    e. all of the above.

67. **In residential properties, forcible entry usually is**
    a. not required.
    b. made through the front door.
    c. accessed through locked windows.
    d. only b and c.
    e. all of the above.

68. **Ventilation is an action at a working fire that**
    a. is used after entry cannot be performed.
    b. when performed properly can have an immediate positive effect.
    c. is used in life saving situations only.
    d. only a and b.
    e. all of the above.

69. **The term "up and over" best describes**
    a. a method to perform ventilation.
    b. a method to perform forcible entry.
    c. a method to perform search and rescue.
    d. all of the above.
    e. none of the above.

70. **Roof ventilation should**
    a. be in conjunction with horizontal ventilation on the upper floors.
    b. occur directly over the fire area.
    c. be an opening in the roof that will draw the fire to that opening.
    d. only b and c.
    e. all of the above.

71. **Hydraulic ventilation is achieved by using hose-lines with a nozzle set on a fog pattern that should cover what percent of the window or door opening through which the smoke will be vented?**
    a. 50 percent.
    b. 60 percent.
    c. 70 percent.
    d. 80 percent.
    e. 90 percent.

72. **Positive-pressure ventilation is accomplished by**
   a. placing fans to push the smoke or toxic fumes from a building.
   b. using fog nozzles to push smoke or toxic fumes from a building.
   c. placing fans to pull the smoke or toxic fumes from a building.
   d. only a and b above.
   e. all of the above.

73. **In regard to a trench cut, factors to consider are**
   a. the time it will take to cut the trench.
   b. the availability of saws to make the cut.
   c. the availability of personnel to complete the task.
   d. only a and c.
   e. all of the above.

74. **At operations involving a trench cut:**
   1) **Once the trench has been opened, the fire side of the roof has been given up for lost.**
   2) **Once opened, the trench location must be monitored from above and below. The hose-line above the cut is usually more effective in preventing the fire from jumping across the trench.**
   a. Both statements are true.
   b. Both statements are false.
   c. Only statement number one is true.
   d. Only statement number two is true.

75. **Hose-line operations into ventilation openings should be**
   a. restricted to roof openings.
   b. restricted to window openings.
   c. restricted to door openings.
   d. all of the above.
   e. none of the above.

76. **Once a fire is under control, suspected areas must be examined for hidden fire. The fire officer must determine**
   a. whether to open a specific area.
   b. where to make openings in a wall or ceiling.
   c. whether a small opening or the entire wall or ceiling area must be opened.
   d. only a and b.
   e. all of the above.

77. 1) Combustible dust lying on a heated surface is subject to ignition due to carbonization of the dust.
     2) Dust explosions usually occur in pairs. The initial explosion may not cause substantial damage but the secondary explosion is usually devastating.
     a. Both statements are true.
     b. Both statements are false.
     c. Only statement number one is true.
     d. Only statement number two is true.

78. In regard to overhauling at an incident scene:
     1) A fireground detail or watch may be needed to ensure that the fire has been fully extinguished.
     2) It is permissible to allow occupants and owners to re-enter the fire building during active operations to collect valuable items such as money, jewelry, or medicine.
     a. Both statements are true.
     b. Both statements are false.
     c. Only statement number one is true.
     d. Only statement number two is true.

79. The National Fire Protection Association categorizes building construction according to types. How many basic types do they list?
     a. 3.
     b. 4.
     c. 5.
     d. 6.
     e. 7.

80. In regard to spalling it
     a. occurs in steel.
     b. has no effect on the strength of a building.
     c. is a condition that can occur in concrete under some fire conditions.
     d. only a and c.
     e. all of the above.

81. When opening a bar joist roof
     a. take care not to cut through the top chord of the roof truss.
     b. the roof opening should be made adjacent to the bar joist.
     c. heated tar on the roof will make the roof slippery and dangerous for firefighters
     d. only a and c.
     e. all of the above.

82. **Composite walls in ordinary constructed buildings that are built today containing both brick and block usually are interconnected by**
    a. metal trusses that are laid atop corresponding courses.
    b. common wall ties that are strips of metal laid between the courses.
    c. steel I-beams.
    d. only a and b.
    e. all of the above.

83. **Typically lumber comes in standard lengths:**
    a. 12-foot long.
    b. 16-foot long.
    c. 20-foot long.
    d. 24-foot long.
    e. none of the above.

84. **A cockloft is**
    a. the space under the roof in a peaked building.
    b. the space under the roof in a flat-roofed building
    c. the crawl space under living quarters.
    d. all of the above.
    e. none of the above.

85. **Any door openings in a fire wall**
    a. must contain sprinklers.
    b. must contain self-closing fire doors.
    c. requires that the fire doors must have the same fire rating as the fire wall.
    d. only b and c.
    e. all of the above.

86. **In regard to heavy timber construction**
    a. columns must be a minimum of 8 inches thick.
    b. girders that span the distance between the columns must be a minimum of 6 inches thick.
    c. the thickness of the floor must be a minimum of 3 inches.
    d. only a and b.
    e. all of the above.

87. **In regard to the heavy timber building, serious problems can develop when a building sits vacant for many years. These include**
    a. roof leaks causing the wooden roof planks to rot.
    b. deterioration of masonry walls due to water infiltration.
    c. weakened walls that can collapse.
    d. only a and b.
    e. all of the above.

88. **Whether a fire can be stopped in a heavy timber building will depend upon**
    a. available resources.
    b. available water supply.
    c. the time of day.
    d. only a and b.
    e. all of the above.

89. **The characteristics of plank and beam frame construction are**
    a. typically uses tongue and groove planks that are a minimum 2-inches thick.
    b. the flammable interior finishes can allow a rapid fire spread.
    c. the beams are often boards laminated together to form large beams.
    d. only b and c.
    e. all of the above.

90. **Wood deterioration can occur due to**
    a. wood-boring insects.
    b. rot or fungus.
    c. water-based paints.
    d. only a and b.
    e. all of the above.

91. **The two types of wood truss construction are**
    a. lightweight truss and bar joist.
    b. bar joist and wooden "I" beams.
    c. lightweight truss and timber truss.
    d. lightweight truss and heavyweight truss.
    e. none of the above.

92. **The interconnection of adjoining truss is called**
    a. cross connecting.
    b. bridging.
    c. parallel chording.
    d. interwebbing.
    e. none of the above.

93. **If no preplan of a building exists and a determination is being made at the scene of an emergency, some indicators that a timber truss is supporting the roof include**
    a. occupancy of the building.
    b. a large open area under the roof.
    c. the shape of the roof.
    d. only b and c.
    e. all of the above.

94.   **Timber trusses can span areas of 100 feet, and be spaced 20 feet on center. Failure of one truss with this spacing can create a roof opening as great as:**
   a.   20 feet by 20 feet.
   b.   20 feet by 40 feet.
   c.   20 feet by 100 feet.
   d.   40 feet by 100 feet.
   e.   none of the above.

95.   **Common types of lightweight truss used in residential construction are the triangular and parallel chord trusses with sheet metal surface fasteners. These fasteners are often referred to as**
   a.   gusset plates.
   b.   gang nailers.
   c.   staple plates.
   d.   only a and c.
   e.   all of the above.

96.   **In regard to wooden "I" beams:**
   a.   They can span wide distances in floor or roof systems.
   b.   Their installation never affects their stability.
   c.   Their web may consist of plywood or oriented strand board.
   d.   Only a and c.
   e.   All of the above.

97.   **Failure of a connector or any part of a truss can cause**
   a.   web members to shift.
   b.   the entire truss to fail.
   c.   multiple truss failures to occur.
   d.   only a and b.
   e.   all of the above.

98.   **To minimize the potential problem of a fire being able to spread throughout large void spaces, most codes have limits on the size of these spaces. This is a form of compartmentation in the truss area and is called**
   a.   void space reduction.
   b.   draft-stopping.
   c.   full space reduction.
   d.   only a and c.
   e.   all of the above.

99.   **A critical cue and probably the most dangerous situation from the standpoint of firefighter safety is**

    a.   a collapse which has occurred before firefighters have arrived on a scene.

    b.   a fire that has been extinguished before the arrival of the fire department.

    c.   a fire burning in a truss void area.

    d.   only a and b.

    e.   all of the above.

100.   **There are a number of considerations for firefighters when confronted with a fire in a building that uses lightweight building components as structural members. Firefighters should**

    a.   anticipate the interconnection of void spaces.

    b.   be alert to weakened floors that they may fall through.

    c.   consider that when pulling or opening ceilings, they should attempt to do so from doorways.

    d.   only a and b.

    e.   all of the above.

ANSWER KEY Study Guide 2 for Chapters 1 through 5

Question	Answer	Page Reference	Question	Answer	Page reference
1	B	4	26	E	42
2	D	11	27	E	43
3	A	12	28	D	54
4	E	14	29	D	57
5	D	14	30	A	60 – 61
6	E	15	31	E	61
7	C	16	32	C	73
8	C	17	33	B	75-77
9	E	18	34	A	77 – 78
10	A	19	35	B	79
11	A	21	36	A	84
12	A	22	37	D	90
13	D	25	38	E	92
14	A	28	39	E	93
15	E	30	40	D	93 – 94
16	D	39	41	E	100
17	D	40	42	D	100 – 101
18	A	40	43	B	101
19	D	40	44	E	102
20	C	40	45	E	103 – 104
21	D	41	46	E	105
22	A	41	47	E	106
23	E	41	48	C	106
24	D	42	49	E	107
25	A	42	50	E	108

ANSWER KEY Study Guide 2 for Chapters 1 through 5

Question	Answer	Page Reference	Question	Answer	Page Reference
51	E	110	76	E	162
52	E	119	77	A	165
53	E	120	78	C	167 – 168
54	E	122	79	C	172
55	E	123	80	C	173
56	E	124	81	E	176
57	E	126	82	D	178
58	E	134	83	C	180
59	D	134	84	B	182
60	D	136	85	D	183
61	C	137	86	E	184
62	E	142	87	E	184 – 185
63	E	144	88	D	186
64	D	146	89	E	188
65	B	147 – 148	90	D	188
66	D	148	91	C	190
67	B	150	92	B	191
68	B	152	93	E	192
69	A	153	94	D	194
70	E	153	95	E	195
71	E	157	96	D	199
72	A	157	97	E	201
73	E	158 – 159	98	B	201
74	C	160	99	C	202
75	E	161	100	E	204 – 205

Study Guide 3 Chapters 1 through 5

1. **Electrical fires are classified as:**
   a. Class A fires.
   b. Class B fires.
   c. Class C fires.
   d. Class D fires.
   e. Class E fires.

2. **The method of heat transfer that occurs through contact of materials is:**
   a. Conduction.
   b. Convection.
   c. Radiation.
   d. Solar.
   e. All of the above.

3. **A thorough inspection of a building or facility can reveal**
   a. locations where problems could occur.
   b. areas that can threaten the lives of civilians and firefighters.
   c. actions that firefighters can take to mitigate a problem.
   d. all of the above.
   e. only a and b above.

4. 1) **It is not necessary to denote dry chemical or CO_2 extinguishing systems on a preplan since they will be readily recognizable on the arrival of the fire department.**
   2) **Special needs of occupants, e.g., handicapped, infirm, should be noted on a preplan.**
   a. Both statements are true.
   b. Both statements are false.
   c. Only statement number one is true.
   d. Only statement number two is true.

5. 1) **Standpipes can slow down firefighters in placing hose streams onto the fire since the fire department may need to pressurize the system.**
   2) **Compartmentation will assist in containment of a fire.**
   a. Both statements are true.
   b. Both statements are false.
   c. Only statement number one is true.
   d. Only statement number two is true.

6. 1) **Seek community support for plans concerning evacuation of the immediate population around target hazards.**
   2) **Some communities have established strong relationships with industrial plants.**
   a. Both statements are true.
   b. Both statements are false.
   c. Only statement number one is true.
   d. Only statement number two is true.

7. 1) To determine the needed fire flow during preplanning requires the application of a "fire flow formula" to conditions observed during an inspection of the premises.
    2) There are occasions when a newly appointed or relatively inexperienced officer, lacking the expertise of a seasoned officer, must quickly judge the amount of water needed to control a fire effectively.
    a. Both statements are true.
    b. Both statements are false.
    c. Only statement number one is true.
    d. Only statement number two is true.

8.   Accordingly to the National Fire Academy's fire flow formula, if exterior structures are being exposed to fire from the original fire building, what percent of the actual fire flow for the building on fire should be added to provide protection for each side of a building that has exterior exposures?
    a. 10 percent.
    b. 15 percent.
    c. 25 percent.
    d. 50 percent.
    e. None of the above.

9. 1) Since firefighting is an exact science, the use of the National Fire Academy's quick-calculation method for fire flow can determine the exact gallons per minute of water required for full fire control.
    2) It has been found that as the amount of fire involvement reaches a stage where a defensive attack is necessary, the National Fire Academy's needed fire flow formula for water supply will be slightly greater than required.
    a. Both statements are true.
    b. Both statements are false.
    c. Only statement number one is true.
    d. Only statement number two is true.

10. 1) The company officer determines the route and regulates the speed of the apparatus on an emergency response.
    2) The company officer should be cognizant of intersections where they may meet other responding units when on an emergency response.
    a. Both statements are true.
    b. Both statements are false.
    c. Only statement number one is true.
    d. Only statement number two is true.

11. 1) **The chief's knowledge of fire science, strategy, tactics, and construction is important to predict fire spread.**
    2) **The chief's assessment must determine whether the size of the fire and the type of building construction will permit a fire to be controlled.**
    a. Both statements are true.
    b. Both statements are false.
    c. Only statement number one is true.
    d. Only statement number two is true.

12. 1) **The chief must interpret the verbal reports received and compare them to what is observed. This comparison will allow the chief to decide if satisfactory progress is being made.**
    2) **The chief must let it be known what is expected of the company officer at an incident scene. Postincident analyses and informal discussions after an incident can help to rectify any problems that occurred.**
    a. Both statements are true.
    b. Both statements are false.
    c. Only statement number one is true.
    d. Only statement number two is true.

13. 1) **Command presence is easily recognized and easily attained.**
    2) **A good leader knows what needs to be accomplished and gives deliberate orders that are understood easily.**
    a. Both statements are true.
    b. Both statements are false.
    c. Only statement number one is true.
    d. Only statement number two is true.

14. **The shouting of orders by a company officer**
    a. reinforces the importance of the order.
    b. allows everyone to recognize that the company officer is the person responsible for everyone actions.
    c. can denote an unsolvable problem.
    d. none of the above.
    e. all of the above.

15. 1) **Visualizing scenarios allows us to prepare for the eventuality of certain occurrences. It lessens the surprises and allows the Incident Commander more time for other decisions.**

2) **Experience allows insight as to whether the tactics employed will achieve the strategic goals or how we have to adjust our strategy.**

a. Both statements are true.
b. Both statements are false.
c. Only statement number one is true.
d. Only statement number two is true.

16. **A written Incident Action Plan should be considered whenever:**

a. Multiple jurisdictions are involved in the response.
b. The incident will involve multiple operational periods.
c. A number of ICS organizational elements are activated.
d. Hazardous materials are involved in the incident.
e. All of the above.

17. **It is Command's responsibility to:**

a. Assess incident priorities.
b. Determines strategy.
c. Determine tactics when there is no Operations Section Chief.
d. Develop an Incident Action Plan.
e. All of the above.

18. **It is Command's responsibility to:**

a. Develop an appropriate organizational structure.
b. Manage incident resources.
c. Coordinate overall incident activities.
d. Ensure safety of on-scene personnel.
e. All of the above.

19. **It is Command's responsibility to:**

a. Coordinate activities of outside agencies.
b. Authorize release of information to the media.
c. Make every decision on the incident scene.
d. All of the above.
e. Only a and b.

20. **The best solution for handling incident problems is**

a. calling for more resources than necessary to keep the firefighters always ready.
b. waiting to see how an incident develops and then requesting the exact number of resources needed.
c. calling for and receiving an adequate number of resources.
d. only a and b.
e. none of the above.

21. **The Safety Officer**
    a. reports directly to the Incident Commander.
    b. surveys the incident scene for unsafe conditions.
    c. discusses safety issues with division and group supervisors.
    d. only a and b.
    e. all of the above.

22. **Staging at an incident scene can**
    a. be used by units without specific orders.
    b. allow units that are not needed to be quickly returned.
    c. permit the goals of the Incident Commander to be achieved.
    d. only a and b.
    e. all of the above.

23. **An Operations Section Chief**
    a. should be designated when there is a great demand on the Incident Commander's time.
    b. will run the operations portion implemented to mitigate the on-scene emergency.
    c. is usually delegated to someone already operating at the scene.
    d. only a and b.
    e. all of the above.

24. **Planning is responsible for the Situation Unit. The concerns of the Situation Unit are:**
    a. What has happened?
    b. What is currently happening?
    c. What may happen?
    d. All of the above.
    e. None of the above.

25. **The Demobilization Unit is responsible for:**
    a. Disassembling the incident organization.
    b. Developing a plan for the release of the resources committed to an incident.
    c. Assisting in the implementation of a plan for the release of resources committed to an incident.
    d. Only b and c.
    e. All of the above.

26. **Base can be used in an Incident Management System**
    a. as a marshalling area for resources.
    b. as an area where primary support activities are performed.
    c. at a major hazardous materials incident.
    d. only a and b.
    e. all of the above.

27.　　　　**In regard to Command:**
　　a.　It must be initiated at all incident scenes.
　　b.　Everyday practice on minor fires allows for a smooth transition to a major incident.
　　c.　Some systems require that the senior officer must assume Command upon his or her arrival at an incident.
　　d.　Only a and b.
　　e.　All of the above.

28.　　　　**A command post provides the Incident Commander**
　　a.　a stationary position from which to command the incident.
　　b.　a location so everyone can find the Incident Commander easily.
　　c.　a location to assemble staff and other resources.
　　d.　a place where management functions occur.
　　e.　all of the above.

29.　　　　**An initial status report should contain**
　　a.　the address where the fire is located.
　　b.　the number of stories and size and construction of the building.
　　c.　fire conditions on arrival.
　　d.　only a and c.
　　e.　all of the above.

30.　　　　**A good timeframe for ongoing status reports is**
　　a.　every 5 minutes.
　　b.　every 10 minutes.
　　c.　every 15 minutes.
　　d.　every 20 minutes.
　　e.　none of the above.

31.　　　　**A firefighter reporting a situation found at an incident scene not only must describe the problem but**
　　a.　tell everyone on the incident scene the nature of the problem.
　　b.　inform only the personnel in the immediate area.
　　c.　inform the Incident Commander of the problem and potential solutions.
　　d.　only a and b above.
　　e.　all of the above.

32.　　　　**The "Planning P" is used for:**
　　a.　Ensuring accuracy when doing preplans.
　　b.　Developing an incident action plan.
　　c.　A tactical worksheet for an incident.
　　d.　As a pattern for developing tactical decisions.
　　e.　None of the above.

33.   **The classical method of decision making is**
- a.   a very methodical and time consuming process.
- b.   a fast way of making decisions.
- c.   the method used by most fire officers of making decision at an incident.
- d.   only a and c.
- e.   all of the above.

34.   **Which of the following is not a level in the command sequence?**
- a.   Incident priorities.
- b.   Size-up.
- c.   Strategy.
- d.   Communications.
- e.   Tasks.

35.   **In regard to incident priorities:**
- a.   Achieving incident stabilization cannot expose firefighters to undue risk.
- b.   Property damage includes fire damage.
- c.   Property damage includes any resultant damage associated with the firefighting effort.
- d.   Only a and b.
- e.   All of the above.

36.   **Size-up information can come from**
- a.   occupants.
- b.   bystanders.
- c.   placards on buildings.
- d.   only a and b.
- e.   all of the above.

37.   **The chief officer arriving on scene after the company officer has assumed Command has certain advantages while performing size-up:**
- a.   He or she has slightly more time for decisions.
- b.   The life hazard will, most likely, already have been assessed.
- c.   There probably will have been an interior assessment.
- d.   Only a and b.
- e.   All of the above.

38.   **When doing size-up, the area of a building would include**
- a.   the building's layout.
- b.   irregular shaped structures when located in areas of high building density.
- c.   the area involved in fire.
- d.   only a and c.
- e.   all of the above.

39. **Location of the fire will determine**
a. the possible travel of a fire.
b. potential interior exposures.
c. the temperature of what is burning.
d. only a and b.
e. all of the above.

40. **Additional resources can be requested in the form of**
a. extra alarms.
b. strike teams.
c. task forces.
d. only a and b.
e. all of the above.

41. **When a fire department is confronted with spills involving hazardous materials, assistance in handling these materials can be gotten from the:**
a. Red Cross.
b. Environmental Protection Agency.
c. Coast Guard.
d. Only b and c.
e. All of the above.

42. **Some areas restrict the height of structures to be built in their community. This is based on**
a. the fire department's ability to safely protect the people living or working in a building.
b. the longest ladder carried by the fire department.
c. a decision by the chief that the fire department does not wish to fight high-rise fires.
d. only a and b.
e. all of the above.

43. **A nighttime fire**
a. in an office or commercial building normally would present little threat to life.
b. could be a threat to the life safety of firefighters.
c. in residential buildings will tax firefighters the greatest.
d. only b and c.
e. all of the above.

44. **In regard to size-up:**
a. Size-up is not reserved for the Incident Commander only.
b. Each individual operating at an emergency scene must perform it.
c. Size-up is performed only on arrival.
d. Only a and b.
e. All of the above.

45.   **The strategies of rescue, exposures, confinement, extinguishment, and overhaul**

a.   are in order of how these strategies must always be performed.

b.   are in order of implementation after ventilation has been performed.

c.   are in priority order of consideration at an emergency.

d.   only a and b.

e.   all of the above.

46.   **In regard to ventilation and salvage**

a.   ventilation is used initially and salvage is the last strategy accomplished.

b.   both are implemented immediately after "RECEO".

c.   they are initiated as needed during the course of firefighting and not at one set juncture.

d.   only b and c.

e.   all of the above.

47.   **The "Strategy Prompter" is a tool**

a.   that determines strategy only.

b.   that can only be used by chief officers.

c.   to help determine how our Incident Management System should be structured.

d.   only a and b.

e.   all of the above.

48.   **The first hose-line has to protect civilians who still are exiting the building or in need of rescue. This line should be placed**

a.   between them and the fire.

b.   to extinguish the fire immediately if possible.

c.   to protect firefighters attempting rescues.

d.   only a and c.

e.   all of the above.

49.   **1)**   **When operating in unoccupied buildings a hose-line should not be operating on an upper floor until the fire on the lower floors either is controlled, or a hose-line is placed in operation on those floors.**

     2)   **Operating in an unoccupied building on a floor above an uncontrolled fire with no hose-line attacking the fire below places firefighters in a high-risk, high-gain situation.**

a.   Both statements are true.

b.   Both statements are false.

c.   Only statement number one is true.

d.   Only statement number two is true.

50. **A rule of thumb for hose-line pulled up the exterior of a building or up a stairwell is**

a. one length of hose-line for every floor.
b. one length of hose-line for every two floors.
c. one length of hose-line for every three floors.
d. each building is different and there is no set rule.
e. none of the above.

51. **An interrupted water supply can occur due to**

a. pump failure.
b. a burst length of hose-line.
c. an inadequate hydrant system.
d. only a and c.
e. all of the above.

52. **When a strictly defensive fire operation is apparent, the first responding officer must**

a. make a reasonable assumption of what the fire will consume and where it can be stopped.
b. be prepared to write off parts of buildings or even entire buildings.
c. wait for the chief officer to arrive to assess the incident scene.
d. only a and b.
e. all of the above.

53. **Transitional modes of fire attack are**

a. only an offensive attack mode.
b. only a defensive attack mode.
c. only an offensive then a defensive attack mode.
d. only a defensive then an offensive attack mode.
e. a combination of offensive and defensive attack modes.

54. **When going from an offensive to a defensive mode of attack**

a. there is a tendency for firefighters to continue operating hand-held hose-lines.
b. it has been found that hand-held hose-lines place firefighters in danger zones.
c. the defensive attack should be performed with master stream devices.
d. only a and b.
e. all of the above.

55. **National standards recommend a minimum staffing in a career fire department of how many firefighters on an engine company?**

a. 2.
b. 3.
c. 4.
d. 5.
e. 6.

56. **In regard to "confinement" the following points must be considered**

    a.   what is being attempted against what can be accomplished.

    b.   what staffing is available.

    c.   what is the risk to firefighters.

    d.   only a and c.

    e.   all of the above.

57. **Building characteristics can permit extension of fire to other areas of the building through**

    a.   balloon frame construction.

    b.   air shafts.

    c.   unprotected horizontal openings.

    d.   only a and b.

    e.   all of the above.

58. **In regard to water supply**

    a.   an evaluation of the available water supply must be made.

    b.   consider assigning a Logistics Section Chief.

    c.   it is only a consideration on large fires.

    d.   only a and b.

    e.   all of the above.

59. **Closing a nozzle too quickly produces water-hammer which can**

    a.   cause the hose-line to burst.

    b.   damage the apparatus.

    c.   damage water mains.

    d.   only a and b.

    e.   all of the above.

60. **In regard to hose-line advancement:**

    a.   A hose-line that does not advance is ineffective.

    b.   Operating a hose-line from a fixed location rather than advancing it will cause more water damage.

    c.   The hose-line must be able to knock down the fire as it moves through the fire area.

    d.   Only a and c.

    e.   All of the above.

61. **An exterior operation involving the use of multiple master streams requires**

    a.   a large volume of water.

    b.   numerous appliances regardless of the water supply.

    c.   maintenance of adequate nozzle pressure.

    d.   only a and c.

    e.   all of the above.

62. **In regard to search and rescue and the use of window stickers to indicate a child's bedroom, the most accurate statement/s is/are**
    a. they are an excellent source of information for firefighters.
    b. these stickers can be misleading.
    c. all rooms must be thoroughly searched for victims.
    d. only b and c.
    e. all of the above.

63. **In regard to search and rescue, the secondary search should**
    a. cover the entire structure, both on the interior and exterior.
    b. only be used if someone is still reported missing.
    c. use a different crew than the primary search if possible.
    d. only a and c.
    e. all of the above.

64. **In regard to search and rescue, the rescuers when removing victims should**
    a. remove them through the interior of the building if possible.
    b. the initial consideration should be removal via ladders.
    c. realize that removal via ladders can cause some problems with unconscious victims.
    d. only a and c.
    e. all of the above.

65. **The best way to gain proficiency in handling portable ladders is through**
    a. studying the various types of ladders that other fire departments use.
    b. knowing which ladders are carried on which apparatus.
    c. teamwork gained through practice.
    d. all of the above.
    e. none of the above.

66. **The correct positioning of the heel or base of the ladder will ensure that a proper climbing angle is achieved. The desired angle is**
    a. 45 degrees.
    b. 60 degrees.
    c. 75 degrees.
    d. 90 degrees.
    e. none of the above.

67. **Basic rules when climbing a ladder include**
    a. extending your arms.
    b. climb the ladder steadily on the balls of your feet.
    c. use your legs to do the climbing.
    d. only b and c.
    e. all of the above.

68.   **In regard to forcible entry**
- a.   the first rule is to try to open the door.
- b.   observe which way doors open.
- c.   breaking down a door is often the fastest way of entering a property.
- d.   only a and b.
- e.   all of the above.

69.   **Ventilation allows**
- a.   the removal of heated smoke and gases.
- b.   rescue of trapped occupants.
- c.   firefighters to locate and approach the fire with hose-lines quickly.
- d.   only a and b.
- e.   all of the above.

70.   **Firefighters when they hear the sound of breaking glass should**
- a.   look to the sound to see if someone is in trouble.
- b.   slope their shoulders downward to allow any glass striking them to fall away from their body.
- c.   ignore the sound and continue with what they are doing.
- d.   only a and b.
- e.   all of the above.

71.   **Firefighters operating on roofs should**
- a.   work in pairs.
- b.   sound out the roof to ensure stability.
- c.   have two remote ways of egress from the roof.
- d.   only b and c.
- e.   all of the above.

72.   **In addition to the conventional ventilation achieved by opening windows, doors and roofs there are mechanical means of ventilation. They include**
- a.   negative pressure.
- b.   hydraulic.
- c.   positive pressure.
- d.   only a and c.
- e.   all of the above.

73.   **In regard to a trench cut:**
- a.   It can be considered when firefighters are confronted with an uncontrolled cockloft fire.
- b.   It is used to cut off the spread and confine the fire to one part of a roof.
- c.   It is effective on peaked roof structures.
- d.   Only a and b.
- e.   All of the above.

74.　1)　Before taking a power saw to a roof it should be checked to see that the proper blade is in the saw to cut the roof.

　　2)　Before taking a power saw to a roof it should be started when it's taken off of the apparatus and allowed to warm up while being carried up the ladder to the roof so it can be used immediately upon arriving on the roof.

　　a.　Both statements are true.
　　b.　Both statements are false.
　　c.　Only statement number one is true.
　　d.　Only statement number two is true.

75.　**Areas of hidden fire sometimes can be detected by**

　　a.　listening for crackling sounds of fire still burning.
　　b.　touching walls or floors to feel for hot spots.
　　c.　through the use of thermal imaging cameras.
　　d.　only b and c.
　　e.　all of the above.

76.　**Structural overhauling of a building often is dependent on the construction classification. Of the following statements:**

　　1)　Noncombustible buildings require minimal examination, since the fire will not affect them structurally.

　　2)　Frame and ordinary constructed buildings have components that will contribute fuel to the fire: There can be many concealed spaces that need to be checked.

　　a.　Both statements are true.
　　b.　Both statements are false.
　　c.　Only statement number one is true.
　　d.　Only statement number two is true.

77.　**In regard to combustible dusts:**

　　a.　Water will extinguish most dust fires.
　　b.　Moisture will raise the ignition temperature of the dust.
　　c.　Fog nozzles should be used to fight a dust fire.
　　d.　Only a and c.
　　e.　All of the above.

78.　**In regards to incident scene preservation for investigative purposes, important points for suppression crews include:**

　　a.　Did anything seem out of the ordinary?
　　b.　Extinguish the fire without destroying the entire scene of potential evidence.
　　c.　The firefighter's ability to recall where each piece of furniture was located.
　　d.　Remembering what the scene/room looked like when they arrived and once inside where was the fire located.
　　e.　All of the above.

79. **Salvage is the preservation of the structure and its contents from additional damage from**
   a. fire and smoke.
   b. water.
   c. firefighting activities.
   d. only a and b.
   e. all of the above.

80. **A fire-resistive building can contain**
   a. steel.
   b. concrete.
   c. other fire-rated materials.
   d. only a and b.
   e. all of the above.

81. **A fire-resistive building is one**
   a. that cannot be damaged by fire.
   b. in which the structural components will resist the effects of fire for a period of time.
   c. where extensive damage can occur due to smoke or fire.
   d. only b and c.
   e. all of the above.

82. **In regard to collapse when opening a bar joist roof:**
   a. Since bar joist are made of steel they will not collapse.
   b. Roof insulation will hide some of the telltale warning signs of collapse.
   c. It can occur before the roof gives any indication of an impending failure.
   d. Only b and c.
   e. All of the above.

83. **The exterior finish on a wall in ordinary construction can be brick, block, stucco, or a thin covering of mortar that is referred to as**
   a. an ordinary finish.
   b. parging.
   c. veneer.
   d. all of the above.
   e. none of the above.

84. **Common bearing walls or party walls may be found in town houses, garden apartments, strip malls, row houses, and similar types of buildings. These buildings**
   a. share the same bearing wall.
   b. may have floor joist share common wall sockets.
   c. can be prone to fire spread through conduction of fire through wooden beams.
   d. only a and b.
   e. all of the above.

85.   **In regard to firefighting in ordinary constructed buildings:**
 a.   The buildings generally are stable and allow an offensive attack.
 b.   Plaster or plasterboard walls will assist in containing a fire.
 c.   Wood lath will supply fuel to a partition fire.
 d.   Only a and c.
 e.   All of the above.

86.   **In regard to a masonry fire wall:**
 a.   Commonly they are thicker than ordinary walls.
 b.   They must extend above a combustible roof.
 c.   They should remain stable should either side of the wall collapse.
 d.   Only a and b.
 e.   All of the above.

87.   **In regard to heavy timber construction:**
 a.   The interior walls and ceilings usually are not finished.
 b.   There are few concealed spaces.
 c.   The floors are usually 2-inch thick wooden floors.
 d.   Only a and b.
 e.   All of the above.

88.   **Mill buildings present a tremendous fire problem due to the**
 a.   building's contents.
 b.   methods of storage.
 c.   large amount of exposed wood within the buildings.
 d.   only a and b.
 e.   all of the above.

89.   **The characteristics of balloon frame construction are**
 a.   a continuous wall stud from the lowest level through the roof area.
 b.   fire-stopping is plentiful.
 c.   void spaces created by the floor joist at each level.
 d.   only a and c.
 e.   all of the above.

90.   **The most common types of frame construction are**
 a.   heavy timber and balloon.
 b.   platform and balloon.
 c.   platform and log.
 d.   post and beam and balloon.
 e.   plank and beam and balloon.

91.   **Wood can be treated to resist attack from**
 a.   fire.
 b.   moisture.
 c.   insects.
 d.   only b and c.
 e.   all of the above.

92.   **The parts of a truss consist of**
    a.   top chord.
    b.   bottom chord.
    c.   web.
    d.   only a and b.
    e.   all of the above.

93.   **A weakness of all trusses and the major reason for truss failure is**
    a.   a fire will attack the truss on many fronts.
    b.   exposed wooden members will ignite.
    c.   steel will conduct heat to the inner wood surfaces destroying the wood fiber in contact with the steel.
    d.   the attack of the fire on the truss causes web members to shift.
    e.   all of the above.

94.   **Fighting a fire in a timber truss building requires constant monitoring. This can best be accomplished by**
    a.   checking both the interior and exterior of the fire building.
    b.   checking whether runoff water is hot to the touch to determine if water lines are hitting the fire.
    c.   continually pushing hose-lines forward into a building until the fire is extinguished.
    d.   only a and b.
    e.   all of the above.

95.   1)   **Trusses have been constructed with plywood gusset plates. Under fire conditions, plywood has a tendency to delaminate, exposing additional fuel to the fire, thus accelerating collapse.**
    2)   **When lumber chars, it has been found that the char offers some protection to the wood in the form of insulation.**
    a.   Both statements are true.
    b.   Both statements are false.
    c.   Only statement number one is true.
    d.   Only statement number two is true.

96.   **In regard to wooden "I" beams the web consists of**
    a.   plywood.
    b.   oriented strand board.
    c.   steel.
    d.   Only a and b.
    e.   All of the above.

97.   **A parallel chord floor truss assembly within a building creates a situation similar to what type of construction?**
    a.   Log.
    b.   Balloon frame.
    c.   Ordinary.
    d.   Noncombustible.
    e.   Brick.

98.   **Draft-stopping is a form of compartmentation. The main problem is that the draft-stopping is too easily negated. Placing holes in the draft-stopping is referred to as**

a.   service openings.
b.   poke-throughs.
c.   service entrances.
d.   only a and c.
e.   none of the above.

99.   **You can find out whether lightweight components have been used in the construction of a property**

a.   by preplanning and documentation during the construction of the building.
b.   if buildings are visibly marked with placards on the exterior.
c.   by asking those at the scene of a building fire.
d.   only a and b.
e.   all of the above.

100.   **There are a number of considerations for firefighters when confronted with a fire in a building that may use lightweight building components as structural members. They should**

a.   anticipate rapid failure of lightweight assemblies.
b.   recognize that fire attacking sheet metal surface fasteners will destroy their connection to the wood.
c.   realize that roof failure can occur without the warning signs found when operating on a roof constructed of solid wooden beams.
d.   only a and c.
e.   all of the above.

ANSWER KEY Study Guide 3 for Chapters 1 through 5

Question	Answer	Page Reference	Question	Answer	Page Reference
1	C	4	26	E	70
2	A	5	27	E	73
3	D	13	28	E	74
4	D	14	29	E	75
5	D	14	30	B	76
6	A	15	31	C	78
7	A	17	32	B	86
8	C	18	33	A	90
9	D	18	34	D	91
10	A	21	35	E	92
11	A	23	36	E	93
12	A	24	37	E	94
13	D	25	38	E	100
14	C	25	39	D	101
15	A	26	40	E	101
16	E	39 – 40	41	D	102
17	E	54	42	D	104 – 105
18	E	54	43	E	105
19	E	54	44	D	106
20	C	56 – 57	45	C	107
21	E	59 – 60	46	C	107
22	E	61 - 62	47	C	110
23	E	63 – 64	48	E	119
24	D	68	49	C	120
25	D	68	50	C	121

ANSWER KEY Study Guide 3 for Chapters 1 through 5

Question	Answer	Page Reference	Question	Answer	Page Reference
51	E	122	76	D	163
52	D	123	77	E	165
53	E	125	78	E	166
54	E	127	79	E	168
55	C	132	80	E	173
56	E	133	81	D	174
57	E	134	82	D	176
58	D	134-135	83	B	178
59	E	136	84	E	180
60	E	136	85	E	182 – 183
61	D	137	86	E	183 – 184
62	D	141	87	D	184
63	D	143	88	E	185
64	D	145	89	D	186
65	C	147	90	B	188
66	C	148	91	E	189
67	E	148	92	E	190
68	D	150	93	E	192
69	E	151	94	D	194
70	B	153	95	A	198
71	E	154	96	D	198
72	E	157	97	B	201
73	D	158	98	B	202
74	C	161	99	D	202
75	E	162	100	E	205 – 206

Study Guide 4 Chapters 1 through 5
MID-TERM

1.   **Heat conducted by a gas or liquid is through?**
    a.   Conduction.
    b.   Convection.
    c.   Radiation.
    d.   Direct flame contact.
    e.   None of the above.

2.   **1)**   **Hands-on training permits firefighters to better understand how the various units function, allowing an emergency scene to operate smoothly.**
    2)   **Training allows members assigned as a rapid intervention crew to be able to use apparatus on the incident scene to assist in removing trapped firefighters.**
    a.   Both statements are true.
    b.   Both statements are false.
    c.   Only statement number one is true.
    d.   Only statement number two is true.

3.   **A preplan can include a section addressing areas that would pose a threat to firefighters. These factors could include**
    a.   the presence and location of hazardous materials in the building – e.g., asbestos, radioactive material, or PCB's.
    b.   flammable or explosive processes.
    c.   the location of open shaftways or chases.
    d.   only a and b.
    e.   all of the above.

4.   **1)**   **Once specific target hazards are identified, potential incident scene problems can be predicted and solutions to those problems can be formulated.**
    2)   **Solutions to anticipated problems at a target hazard could be addressed by developing standardized procedures for these recurring situations. The implementation of set procedures permits them to be practiced under nonemergency conditions to hone firefighter skills.**
    a.   Both statements are true.
    b.   Both statements are false.
    c.   Only statement number one is true.
    d.   Only statement number two is true.

5. 1) Review of the preplan with the personnel of the affected facility may open their eyes to the distinct possibility of destruction of their facility.
    2) Reviewing a preplan with the personnel of an affected facility can initiate major changes within that facility to prevent a disaster from occurring.
    a. Both statements are true.
    b. Both statements are false.
    c. Only statement number one is true.
    d. Only statement number two is true.

6. Accordingly to the National Fire Academy's fire flow formula, for a fire involving 50 percent of the first floor of a 2-story building that is 30 feet by 40 feet with no exterior exposures, the estimated fire flow would be
    a. 125 gallons per minute.
    b. 250 gallons per minute.
    c. 400 gallons per minute.
    d. 500 gallons per minute.
    e. none of the above.

7. Accordingly to the National Fire Academy's fire flow formula, if an interior exposure becomes involved in fire what amount of fire flow should be added to the original fire flow?
    a. 10 percent of the original fire flow.
    b. 15 percent of the original fire flow.
    c. 25 percent of the original fire flow.
    d. 50 percent of the original fire flow.
    e. none of the above.

8. If the fire flow capability of available resources exceeds the required fire flow, an interior attack on the fire usually can be made. However, before this decision is implemented, the Incident Commander should consider which of the following?
    a. Do existing conditions allow a sufficient degree of safety for the firefighters performing an interior attack?
    b. Are there sufficient firefighters on scene?
    c. Is the fire area accessible?
    d. How many hose-lines and firefighters are needed?
    e. All of the above.

9.   **An important skill for a company officer is motivation. He or she should**
   - a. attempt to find out what drives each individual.
   - b. be personally motivated.
   - c. realize that some individuals are motivated by praise for a job well done.
   - d. realize that some people may be motivated by a pay raise.
   - e. all of the above.

10.   1) **The company officer on arriving at the incident scene must size-up the structure. This observation must include anything out of the ordinary.**
       2) **The company officer is responsible for initial fire investigations as to the cause of fire.**
   - a. Both statements are true.
   - b. Both statements are false.
   - c. Only statement number one is true.
   - d. Only statement number two is true.

11.   1) **Chief officers who delegate assignments to subordinates will find themselves overwhelmed with questions on minor details.**
       2) **Delegation is a major part of leadership.**
   - a. Both statements are true.
   - b. Both statements are false.
   - c. Only statement number one is true.
   - d. Only statement number two is true.

12.   1) **The chief, when assuming the role of the Incident Commander, must ensure that no one takes unnecessary chances that would endanger themselves or others while attempting to control or extinguish a fire.**
       2) **The chief needs an intricate knowledge of the implementation of a command system to handle the large variety of problems that could occur.**
   - a. Both statements are true.
   - b. Both statements are false.
   - c. Only statement number one is true.
   - d. Only statement number two is true.

13.   **Command presence consists of the following traits:**
   - a. Self-discipline.
   - b. A take-charge person.
   - c. Easily persuaded to change their mind.
   - d. Deliberate but not necessarily precise.
   - e. Both a and b.

14. 1)   **A hazardous materials incident may involve a relatively short exposure to firefighters, but can result in long-term disabling complications.**

    2)   **Attempting to accomplish too many tasks with insufficient personnel will achieve the desired goals.**

    a.   Both statements are true.
    b.   Both statements are false.
    c.   Only statement number one is true.
    d.   Only statement number two is true.

15. **The aim of the firefighter life safety initiatives is:**
    a.   Advocate the need for a cultural change.
    b.   Empower all firefighters to stop unsafe practices.
    c.   Develop and champion protocols for response to violent incidents.
    d.   Only a and b.
    e.   All of the above.

16. **At a fire or other incident, success or failure is entrusted to**
    a.   the highest ranking chief officer.
    b.   the Incident Commander.
    c.   the captain of the first-due engine company.
    d.   the second-due chief officer.
    e.   none of the above.

17. **When an Incident Commander gets physically involved in firefighting they**
    a.   will be able to better accomplish assignments.
    b.   may overlook some critical areas.
    c.   have better control of firefighters due to the close contact.
    d.   all of the above.
    e.   none of the above.

18. **The basic assignments for a building fire usually will include**
    a.   one company on fire attack.
    b.   one company assigned to search and rescue.
    c.   one company performing ventilation.
    d.   a rapid intervention crew.
    e.   all of the above.

19. **Command must evaluate decisions made prior to his or her arrival and decide**
    a.   whether to continue with the plan already in place.
    b.   whether to institute a new plan.
    c.   whether the current plan will meet the necessary goals.
    d.   only a and b.
    e.   all of the above.

20.   **The Safety Officer should**
   a.   report to the Incident Commander on arrival at the incident scene.
   b.   give orders to divisions and groups on strategy and tactics.
   c.   present safety concerns to Command for action.
   d.   only a and c above.
   e.   all of the above.

21.   **The position of Intelligence/Investigations Officer is normally staffed by:**
   a.   A fire officer.
   b.   A wildland chief officer.
   c.   A police official.
   d.   A local politician.
   e.   All of the above.

22.   **The correct title for Operations is:**
   a.   Command.
   b.   Section Chief.
   c.   Director.
   d.   Leader.
   e.   Manager.

23.   **The following is a responsibility of a Branch Director:**
   a.   Keeps his or her supervisor informed of the status in the Branch's area of responsibility.
   b.   Assigns specific tasks to Divisions, Groups, or Units within the Branch.
   c.   Resolves logistical problems associated with the Units deployed in the Branch.
   d.   All of the above.
   e.   None of the above.

24.   1)   **The Planning Section Chief may interact with technical specialists who can provide insight on an operation.**
       2)   **Technical specialists may be from the private sector.**
   a.   Both statements are true.
   b.   Both statements are false.
   c.   Only statement number one is true.
   d.   Only statement number two is true.

25.   **The Finance/Administration Section Chief's position can be implemented**
   a.   at a major incident.
   b.   if no major cost recovery is evident.
   c.   to ensure that anticipated costs are properly documented.
   d.   only a and c.
   e.   all of the above.

26.   **The absence of Command**
- a.   usually is not noticeable in large fire departments.
- b.   can streamline many fire scene operations.
- c.   allows for indecision and duplication of effort.
- d.   frees up personnel to do other things.
- e.   none of the above.

27.   **1)**   **As higher-ranking officers arrive at an emergency, the incident may require a higher level of experience, necessitating a transfer of Command to the higher-ranking officer.**
     2)   **Whether the higher-ranking officer assumes Command or allows it to remain with the present officer, he or she still assumes the responsibility of the scene.**
- a.   Both statements are true.
- b.   Both statements are false.
- c.   Only statement number one is true.
- d.   Only statement number two is true.

28.   **In regard to transfer of Command:**
- a.   A transfer of information must occur before a good transfer of Command takes place.
- b.   A higher-ranking officer can assume Command and keep the officer who has been relieved at the command post.
- c.   It is unfair to allow a low ranking officer to remain in Command of a very complex incident.
- d.   Only a and b.
- e.   All of the above.

29.   **A command post provides the Incident Commander**
- a.   a location to collect and disseminate information.
- b.   a place where decisions are made.
- c.   a place with lights for nighttime operations.
- d.   only a and b.
- e.   all of the above.

30.   **An initial status report should review**
- a.   the basic points of size-up.
- b.   the life safety of the occupants.
- c.   potential areas that will require overhaul.
- d.   only a and b.
- e.   all of the above.

31.   **Ongoing status reports**
- a.   allow the Incident Commander to review his or her size-up.
- b.   allow the Incident Commander to determine if the present strategy and tactics are accomplishing the intended goals.
- c.   are used to give initial conditions found at an incident.
- d.   only a and b.
- e.   all of the above.

32. 1) **A good communicator can read the body language of a person giving or receiving a message when face-to-face with that individual. These nonverbal messages can be expressions or gestures.**
     2) **The repeating back of an order by the firefighter receiving the order is unnecessary. Further questions can clarify the order if it has not been fully understood.**
     a. Both statements are true.
     b. Both statements are false.
     c. Only statement number one is true.
     d. Only statement number two is true.

33. **A written incident action plan**
     a. contains work assignments.
     b. facilitates dissemination of critical information.
     c. contains incident objectives.
     d. contains the response strategy.
     e. all of the above.

34. **The Radio Communications Plan is ICS Form number?**
     a. 201.
     b. 202.
     c. 203.
     d. 204.
     e. 205.

35. **Cue-based decision making uses the fire officer's knowledge gained through**
     a. personal experience.
     b. training exercises.
     c. incident scene responses.
     d. only a and c.
     e. all of the above.

36. **Incident priorities are**
     a. life safety.
     b. property conservation.
     c. incident stabilization.
     d. only a and c.
     e. all of the above.

37. **Minimizing property loss benefits**
     a. the building owner.
     b. the occupants.
     c. the community.
     d. the fire department.
     e. all of the above.

38.   **On arrival at the fire scene the first arriving officer should**
- a.   give orders to all responding units immediately on arrival.
- b.   never assume Command until other units arrive.
- c.   take a 360-degree walk-around the fire building or incident.
- d.   only a and c.
- e.   all of the above.

39.   **"Wallace Was Hot" denotes**
- a.   a method of developing strategies.
- b.   implementation of tactics.
- c.   size-up factors.
- d.   only a and b.
- e.   all of the above.

40.   **The area of a fire building can be determined by**
- a.   asking a bystander.
- b.   by referencing a preplan.
- c.   through guesswork.
- d.   only a and b.
- e.   all of the above.

41.   **When dealing with a large structure or a complex of interconnected buildings**
- a.   it is helpful to glean information from the preplan.
- b.   an engineer from the complex present at the command post can answer questions.
- c.   there is really no need for preplan information.
- d.   only a and b.
- e.   all of the above.

42.   1)   **To achieve incident control it is important to consider which apparatus can be used for rescue, fire attack, or protection of exposures.**
      2)   **The quantity of water needed to control the fire and the ability of the apparatus pumps to deliver that amount of water must be considered.**
- a.   Both statements are true.
- b.   Both statements are false.
- c.   Only statement number one is true.
- d.   Only statement number two is true.

43.   **One major reason for conflagrations due to poor construction methods is**
- a.   lack of an Incident Management System.
- b.   wood shingle roofs.
- c.   poor fire protection.
- d.   lack of water supply.
- e.   none of the above.

44.   **When considering the size-up factor "weather:"**
- a.   Freezing weather can increase response times.
- b.   Apparatus will be slowed if ice or snow is on the ground.
- c.   Slippery surfaces can slow and injure firefighters.
- d.   Fire escapes and exterior stairs may be slippery and dangerous.
- e.   All of the above.

45.   **The size-up factor "special matters" can include**
- a.   topography.
- b.   urban interface at a wildland fire.
- c.   elevated highways, railways, or bridges.
- d.   only b and c.
- e.   all of the above.

46.   **The size-up factor "height" should recognize that**
- a.   a building of greater height can be exposed on an upper floor from radiant heat.
- b.   adjoining structures of greater height may have windows above the roofline that can be exposed should a fire break through the roof.
- c.   building height is an overstated and should be ignored.
- d.   only a and b.
- e.   all of the above.

47.   **A daytime fire in an office or commercial building**
- a.   presents a threat to the lives of employees working in the building.
- b.   allows early discovery of the fire.
- c.   has no effect in discovering the fire.
- d.   only a and b.
- e.   all of the above.

48.   **The systematic deployment of strategy should be considered a tool and used along with**
- a.   knowledge.
- b.   experience.
- c.   training.
- d.   only a and b.
- e.   all of the above.

49.   **The acronym "RECEO – VS" refers to**
- a.   rescue, exposures, containment, extension, overhaul, ventilation, saving life.
- b.   rescue, extinguishment, containment, exposures, overhaul, ventilation, salvage.
- c.   rescue, exposures, confinement, extinguishment, overhaul, ventilation, salvage.
- d.   rescue, extinguishment, control, exposures, overhaul, ventilation, salvage.
- e.   none of the above.

50. When considering confinement of a fire the Incident Commander should

    a. try to confine the fire to as small an area possible.

    b. predict what problems must be overcome to achieve confinement.

    c. order fire doors to be shut and have companies supply sprinkler systems.

    d. only a and b.

    e. all of the above.

51. The reason that tactics need to be measurable and specific

    a. is to determine when they have been accomplished.

    b. so that the Incident Commander can be methodical in giving orders.

    c. because the chief is the Incident Commander.

    d. only a and b.

    e. all of the above.

52. 1. Crew Resource Management recognizes that humans behave predictably, and if certain behaviors are learned and practiced, operations can be made significantly safer.

    2. Crew Resource Management uses training to ensure that the members of the flight crew are bold and assertive.

    a. Both statements are true

    b. Both statements are false

    c. Only statement number one is true

    d. Only statement number two is true

53. Gaining control of the stairs

    a. is the first step in gaining control of the building.

    b. ensures that protected stairs can be used by firefighters entering the structure and civilians leaving the building.

    c. through proper hose-line placement can reduce the heat of the fire extending to the upper floors via the stairs.

    d. only a and b.

    e. all of the above.

54. When stretching hose-line for interior operations, company officers must keep in mind that when more than two hose-lines are pulled through the same doorways or up the same stairs

    a. a maximum amount of water can then be delivered.

    b. it becomes very difficult for any of the hose-lines to advance.

    c. someone must determine which hose-line belongs to which engine company.

    d. only a and b.

    e. all of the above.

55. **The nozzle on a hose-line should be opened**
a. when the hose-line reaches the fire area.
b. when the hose-line reaches the front door.
c. when the next engine company arrives on the scene and establishes a continuous water supply.
d. only a and b.
e. all of the above.

56. **1) When using a fog stream to ventilate a room, it works best when the entire pattern is discharged through the window or door opening to the exterior of the building, with allowance for some opening around the stream.**
2) Ventilation will increase the temperature of the room allowing firefighters to enter and complete the extinguishment of the fire.
a. Both statements are true.
b. Both statements are false.
c. Only statement number one is true.
d. Only statement number two is true.

57. **Exterior fire operations must consider**
a. the correct appliance to deliver the hose stream.
b. stream placement.
c. the maximum effectiveness of the hose stream.
d. only a and b.
e. all of the above.

58. **A structure fire may have control of the better part of a building, yet rescues still can be accomplished on the interior under the protection of a hose-line. On completion of the rescues, all units will be removed from the structure and implement an exterior attack. In this particular instance the mode of attack employed is referred to as**
a. offensive.
b. defensive.
c. offensive/defensive.
d. defensive/offensive.
e. nonintervention.

59. **Locating a fire by the initial arriving units can be assisted by**
a. the initial information given to dispatch.
b. visual indicators such as fire showing from a window.
c. information from occupants or bystanders.
d. only b and c.
e. all of the above.

60.   **In determining strategy the Incident Commander must**

    a.   make a reasonable assumption of the size of the area that the fire can be confined to.

    b.   use size-up to assess the current situation.

    c.   use his or her experience to forecast anticipated problems and occurrences.

    d.   only b and c.

    e.   all of the above.

61.   **When predicting fire conditions at an incident scene and considering "confinement" one must take into account**

    a.   whether the building is protected by a sprinkler system.

    b.   if fire walls and fire doors will assist in containment.

    c.   if the building is compartmentalized.

    d.   only a and b.

    e.   all of the above.

62.   **When a company is advancing a hose-line, the best position for the company officer is behind the nozzle-person. This allows the company officer to**

    a.   observe fire conditions.

    b.   observe the effect the stream is having on the fire.

    c.   supervise an inexperienced firefighter.

    d.   order the line to be advanced as knockdown is achieved.

    e.   all of the above.

63.   **In regard to water supply involving master streams**

    a.   use of large size tips should be considered.

    b.   shorter hose-line leads will reduce friction loss.

    c.   a limited water supply will make fire control difficult.

    d.   only a and c.

    e.   all of the above.

64.   **Building standpipe systems that provide 2½-inch hose-line connections and are intended for use by firefighters for full-scale firefighting are classified as:**

    a.   Class A.

    b.   Class B.

    c.   Class C.

    d.   Class 1.

    e.   Class 2.

65.   **In regard to search and rescue, the search is broken down into**

    a.   two phases: initial and ongoing.

    b.   two phases: primary and secondary.

    c.   three phases: primary, ongoing, and final.

    d.   three phases: initial, secondary, and final.

    e.   none of the above.

66. **In regard to search and rescue, the exterior search should**
   a. check under debris.
   b. thoroughly check shrubbery.
   c. be used only when someone is unaccounted for.
   d. only a and b.
   e. all of the above.

67. **In regard to search and rescue, the rescuers when removing victims should**
   a. ensure that the head of the ladder is even with or slightly below the windowsill.
   b. realize that where many rescues are required, ladder operations tend to be slow.
   c. realize that safety may need to be ignored if victims are missing.
   d. only a and b.
   e. all of the above.

68. **1) The assignment to be accomplished with a ladder will determine where the top or head of the ladder should be located.**

   2) The positioning of a ladder for different operations (e.g., rescue, entering a window, etc.) should be figured out at an incident scene; prior considerations will not be beneficial on ladder positioning since each situation will be different.
   a. Both statements are true.
   b. Both statements are false.
   c. Only statement number one is true.
   d. Only statement number two is true.

69. **The correct positioning of the heel or base of the ladder will ensure that a proper climbing angle is achieved. This can be accomplished by placing the heel or base of the ladder _____ of the distance of the height from where the ladder will rest against the building:**
   a. one-half.
   b. one-third.
   c. one-fourth.
   d. one-fifth.
   e. none of the above.

70. **A consideration for ladder placement is**
   a. that two ladders are placed on each side of a fire building.
   b. to ensure that firefighters operating on a truck company use the portable ladders that they carry on their apparatus.
   c. to provide firefighters operating within a structure another means of egress.
   d. only a and c.
   e. all of the above.

71.   **In regard to forcible entry:**
- a.   In residential buildings doors usually open inward.
- b.   In public assembly buildings exit doors should open outward.
- c.   Door hinges can indicate the way a door opens.
- d.   Only a and b.
- e.   All of the above.

72.   **Ventilation when used properly**
- a.   minimizes fire damage.
- b.   improves interior conditions.
- c.   creates better visibility for search and rescue operations.
- d.   can prevent a smoke explosion or backdraft situation.
- e.   all of the above.

73.   **An excellent method of venting the upper areas of a fire building is**
- a.   using skylights or scuttles that are built into a roof structure.
- b.   breaking and opening the front and rear doors of a fire building.
- c.   opening windows on the lower floors.
- d.   only a and c.
- e.   all of the above.

74.   **In regard to roof ventilation at fires in private dwellings:**
- a.   The roof should be opened on arrival of the first-due truck company.
- b.   The roof should be opened immediately after the windows are opened on the upper floors.
- c.   Normally the roof does not need to be opened.
- d.   Only a and b.
- e.   All of the above.

75.   **In regard to a trench cut:**
- a.   It should not be attempted until a ventilation hole of sufficient size is placed directly above the fire area.
- b.   It is not effective on garden apartments.
- c.   It should be located in the expected path of fire travel.
- d.   Only a and c.
- e.   All of the above.

76. In regard to cutting of a trench cut:
    1. **Observation holes should be cut.**
    2. **Cut an adequate ventilation hole directly over the fire.**
    3. **The trench should consist of two parallel cuts 30 to 36 inches apart.**
    4. **Perpendicular cuts are made between the parallel cuts to assist in opening the trench.**
    The correct order of the above operation should be
    a. 1,2,3,4.
    b. 2,3,4,1.
    c. 2,1,3,4.
    d. 4,3,2,1.
    e. none of the above.

77. **Directing a hose-line into any roof opening can**
    a. defeat the purpose of the vent hole.
    b. drive the heat of the fire downward.
    c. endanger firefighters working below.
    d. only b and c.
    e. all of the above.

78. **By using a tool to sound out or probe a roof a firefighter**
    a. can locate the roof surface.
    b. can ensure roof stability.
    c. can check for any light or air shafts.
    d. only a and b.
    e. all of the above.

79. **In regard to overhauling at a fire scene:**
    a. Articles on the top of bureaus can be gently swept into the top drawers.
    b. Burnt clothing can be set aside inside a dwelling so they can be salvaged by the homeowner.
    c. Mattresses should be removed to the exterior.
    d. Only a and c.
    e. All of the above.

80. **Many injuries occur during overhauling due to firefighters**
    a. being tired.
    b. removing protective clothing.
    c. letting their guard down.
    d. all of the above.
    e. none of the above.

81.   **In regard to the properties of steel:**
- a.   It is a ready conductor of heat.
- b.   When heated it will expand about 9½ inches per 100 feet.
- c.   A rule of thumb for firefighters for steel expansion is 1 inch for every 10 feet.
- d.   Due to its high conductivity of heat it must be protected to resist the heat of a fire.
- e.   All of the above.

82.   **The person most qualified to assist firefighters in assessing the strength of concrete at a construction site is the**
- a.   fire chief.
- b.   cement finisher.
- c.   codes inspector.
- d.   design engineer.
- e.   all of the above.

83.   **Noncombustible/limited-combustible constructed buildings**
- a.   use materials that do not contribute to the development or spread of fire.
- b.   offer little resistance to fire.
- c.   will not collapse due to the type of construction.
- d.   only a and b.
- e.   all of the above.

84.   **Safety tips for operating on any roof where the stability is suspect can include**
- a.   a firefighter using a safety belt attached to a main ladder or aerial platform.
- b.   laying a portable ladder on the roof to span the distance between roof joist or rafters.
- c.   using a large sized fire axe for chopping to reduce the number of blows needed to create an opening in the roof.
- d.   only a and b.
- e.   all of the above.

85.   **Operating on or opening a roof of unprotected steel during a building fire should be**
- a.   no problem if accomplished within 20 minutes of arrival on the scene.
- b.   considered dangerous.
- c.   strongly considered as the best means of ventilation.
- d.   only b and c.
- e.   all of the above.

86. **A fire-cut on wood floor joist is intended to**
    a. allow the floor to collapse without pushing the masonry wall outward.
    b. allow the floor to collapse and push the masonry wall outward.
    c. keep the floor from collapsing.
    d. none of the above.
    e. all of the above.

87. **Heavy timber construction is often referred to as**
    a. balloon frame buildings.
    b. plank and beam buildings.
    c. log cabins.
    d. mill buildings.
    e. none of the above.

88. **The roof of a heavy timber constructed building can be supported by columns or fabricated of heavy timber truss. The common truss configurations are**
    a. flat and constructed of parallel chords.
    b. triangular shaped.
    c. bowstring shaped.
    d. only a and b.
    e. all of the above.

89. **Modifications made to heavy timber buildings that can create problems for firefighters include**
    a. concealed spaces and many voids.
    b. pipe chases that afford a ready means of fire to extend to the upper levels.
    c. additional draft curtains.
    d. only a and b.
    e. all of the above.

90. **In regard to heavy timber buildings:**
    a. Once ignited they generate massive amounts of heat.
    b. Fires in these buildings are difficult to control and can severely threaten exposed buildings.
    c. They contain few concealed spaces unless modified.
    d. Only a and b.
    e. All of the above.

91. **If spans of over 20 feet is desired in balloon frame and platform frame construction**
    a. then interior walls may need to act as bearing walls.
    b. lightweight trusses and wooden "I" beams may be used.
    c. spans over 20 feet are not permitted in these types of construction.
    d. only a and b.
    e. all of the above.

92. **Trusses are distinguished by the size of their framing members. To be a timber truss, the minimum size of these members must be**
    a. 2 inch by 4 inch.
    b. 4 inch by 4 inch.
    c. 4 inch by 6 inch.
    d. 4 inch by 8 inch.
    e. none of the above.

93. **The more common method of attachment of chord and web members in a timber truss is**
    a. the use of nails and screws only.
    b. wooden dowels and mortise joints.
    c. steel gusset plates.
    d. wooden plates.
    e. none of the above.

94. **Timber truss roofs fail as do any roof. Factors or indicators associated with their collapse are**
    a. a heavy body of fire in the truss area.
    b. overloading of supporting members.
    c. installation of loads either atop the roof or beneath it.
    d. only a and b.
    e. all of the above.

95. 1) **Oriented strand board is widely used for: construction sheathing, web materials for wooden "I" beams and other applications.**
    2) **Oriented strand board achieves the same advantages of cross-laminated veneers as in plywood.**
    a. Both statements are true.
    b. Both statements are false.
    c. Only statement number one is true.
    d. Only statement number two is true.

96. **To increase the carrying capacity of a truss:**
    a. Increase the number of web members.
    b. Place trusses side by side.
    c. Increase the size of the truss members.
    d. Only a and b.
    e. All of the above.

97. 1) **When solid joist lumber is used in a floor or roof system, a void area is created between each joist.**
    2) **When parallel chord trusses are used in a floor or roof system, one large void area is created.**
    a. Both statements are true.
    b. Both statements are false.
    c. Only statement number one is true.
    d. Only statement number two is true.

98. **In buildings containing lightweight building components a void space fire indicates a dangerous condition since**

    a. lightweight components are especially vulnerable to fire.

    b. lightweight components can fail quite readily under fire attack.

    c. the void space fire may not be recognized until the collapse of lightweight components has occurred.

    d. only a and b.

    e. all of the above.

99. **How should firefighters react when confronted with a fire in a structure containing lightweight components? They should**

    a. consider their incident priorities.

    b. realize that the potential for firefighters becoming involved in a collapse is increased.

    c. give their greatest effort to afford the best chance of survival to a trapped civilian.

    d. only b and c.

    e. all of the above.

100. **A buildup of ice, snow, or water on a roof that is constructed of lightweight components**

    a. could precipitate an early collapse of the roof.

    b. will have no effect since the roof is built to hold heavy loads.

    c. will not occur if the structure is properly built.

    d. only b and c.

    e. all of the above.

ANSWER KEY Study Guide 4 for Chapters 1 through 5

Question	Answer	Page Reference	Question	Answer	Page reference
1	B	5	26	C	73
2	A	11	27	A	73
3	E	14	28	E	73
4	A	14	29	E	75
5	A	16	30	D	75
6	B	17	31	D	77
7	E	18	32	C	78
8	E	18 – 19	33	E	83
9	E	20	34	E	85
10	A	22	35	E	90
11	D	23	36	E	91 – 92
12	A	24 – 25	37	E	92
13	E	25	38	C	93
14	C	26	39	C	99
15	E	28	40	B	100
16	B	56	41	D	101
17	B	57	42	A	101
18	E	57	43	B	102
19	E	59	44	E	103 – 104
20	D	59 – 60	45	E	104
21	C	61	46	D	104
22	B	63	47	D	105
23	D	65	48	E	106
24	A	68	49	C	107
25	D	70	50	E	108

ANSWER KEY Study Guide 4 for Chapters 1 through 5

Question	Answer	Page Reference	Question	Answer	Page Reference
51	A	108	76	C	160
52	A	112	77	E	161
53	E	119	78	E	161
54	B	121	79	D	163 – 164
55	A	121	80	D	167
56	C	122	81	E	173
57	E	124	82	D	174
58	C	125	83	D	174
59	E	133	84	D	176
60	E	133 – 134	85	B	177
61	E	134	86	A	181
62	E	136	87	D	183
63	E	137	88	E	184
64	D	139	89	D	184
65	B	142	90	E	185
66	D	143	91	D	188
67	D	145	92	C	190
68	C	147	93	C	192
69	C	148	94	E	194
70	C	149	95	A	197
71	E	150	96	E	201
72	E	152	97	A	201
73	A	153	98	E	202
74	C	157	99	E	202 – 203
75	D	158	100	A	206

Study Guide 5 Chapters 6 through 11

1. **Most building collapses**
   a. occur on the interior of a structure.
   b. occur on the exterior of a structure.
   c. are spectacular events.
   d. only b and c.
   e. all of the above.

2. **The three basic types of wall collapse are**
   a. 90-degree, inward-outward, explosive.
   b. 90-degree, curtain, explosive.
   c. 90-degree, inward-outward, curtain.
   d. curtain, inward-outward, explosive.
   e. none of the above.

3. **In regard to a collapse zone:**
   a. Firefighters on the roof of another building are considered out of a collapse zone.
   b. Firefighters operating on elevating platforms are considered out of the collapse zone.
   c. Master streams should be placed in a flanking position outside of the collapse zone.
   d. Only a and b.
   e. All of the above.

4. **As fire attacks a building it will destroy its integrity. What structural component more than any other component can contain a multitude of collapse indicators?**
   a. The exterior walls of a building.
   b. The roof of a building.
   c. The doors and windows of a building.
   d. Only a and c.
   e. All of the above.

5. **As failure occurs on the interior of a building, the failing structural members can exert an outward pressure on the exterior wall. This may be visible on the exterior by**
   a. walls bulging.
   b. walls leaning outward.
   c. failure of part of an exterior wall.
   d. only a and b.
   e. all of the above.

6.　　**Firefighters finding wall spreaders in a structure involved in fire**
a.　do not have to worry since the wall spreaders provide additional support for the wall.
b.　usually don't know why the spreaders were installed.
c.　must assume the spreaders were installed to rectify a problem with the wall.
d.　only b and c.
e.　all of the above.

7.　　**Rainwater or water from firefighting operations can be retained on flat roofs that have clogged or frozen drains. This could indicate that**
a.　a large buildup of water can be dangerous.
b.　some water on the roofs of exposed buildings can be beneficial in preventing fires from flying embers.
c.　a tool may be needed to unclog the roof drains of debris.
d.　only a and b.
e.　all of the above.

8.　　**Firefighters may find that a building undergoing demolition**
a.　may have some safety features removed.
b.　that fire stops may be nonexistent.
c.　that demolition experts will weaken critical areas.
d.　only a and b.
e.　all of the above.

9.　　**The live load of a building would include all of the below except:**
a.　Desks.
b.　Furniture.
c.　Machinery.
d.　Kitchen utensils.
e.　Plaster walls.

10.　　**The ability to rescue victims trapped in a collapsed building is dependent upon**
a.　the type of collapse.
b.　the extent of the area involved.
c.　how many floors of a structure have collapsed.
d.　the physical health of a trapped victim.
e.　all of the above.

11.　　**The scene of a collapse that involves rescue operations and many injuries may require**
a.　a staging area and Staging Area Manager.
b.　a Medical Group Supervisor.
c.　monitoring the fatigue level of the rescuers.
d.　only b and c.
e.　all of the above.

12. **When confronted with a collapsed building and the possible need for rescues within the building the Incident Commander should consider requesting**
  a. a heavy rescue unit.
  b. basic and advanced life support medical units.
  c. hospital surgical teams to assist at the scene.
  d. only a and b.
  e. all of the above.

13. **Firefighters attempting to locate victims in a collapsed building should**
  a. do a close inspection by walking over the collapsed area.
  b. interview any victims who have been rescued for information.
  c. surface victims should not be rescued first since others may need immediate help.
  d. only a and b.
  e. all of the above.

14. **When confronted with a collapsed building and a need to locate victims the Incident Commander should ensure that**
  a. someone is assigned to interview bystanders for information.
  b. information is gathered, since people reported trapped may have been rescued prior to the fire department's arrival.
  c. operations proceed slowly due to the possibility of a secondary collapse.
  d. only b and c.
  e. all of the above.

15. **At a building collapse incident**
  a. freestanding walls can be stabilized temporarily by the use of shoring.
  b. shoring is meant to prevent further movement of a floor or wall.
  c. shoring should not be an attempt to restore the area to its original position.
  d. only a and c.
  e. all of the above.

16. **A fire department Safety Officer should**
  a. be highly motivated.
  b. have good managerial skills.
  c. be able to determine safe or unsafe acts at an incident scene.
  d. command the respect of supervisors, peers and subordinates alike.
  e. all of the above.

17. 1) **Carbon tetrachloride previously used in fire extinguishers is a non-carcinogen.**
     2) **Creosote is a known carcinogen that causes cancer of the face, lung, neck, penis, scrotum and skin.**
     a. Both statements are true
     b. Both statements are false.
     c. Only statement number one is true.
     d. Only statement number two is true.

18. **The use of a Safety Officer at an incident scene**
     a. strengthens and supports the incident command organization.
     b. provides specialized knowledge.
     c. ensures that recommended safety procedures are being followed.
     d. only b and c.
     e. all of the above.

19. **The Safety Officer can gain respect and become more effective if they**
     a. act strongly and authoritatively to every unsafe act, even minor ones, by correcting them immediately.
     b. ignore minor unsafe acts since the incident scene must be controlled.
     c. discuss any observed unsafe situations that could lead to an injury with the Division/Group Supervisors.
     d. only a and c.
     e. all of the above.

20. **In regard to the Safety Officer's monitoring of rehab:**
     a. Emergency medical services personnel must evaluate the firefighters.
     b. The firefighter can return to firefighting after being monitored when he or she is ready.
     c. Emergency medical personnel should decide whether a firefighter can return to firefighting.
     d. Only a and c.
     e. All of the above.

21. **A good practice is to reserve the front of the fire building for which apparatus?**
     a. The heavy rescue.
     b. The first-due engine.
     c. The truck company.
     d. The first-due chief.
     e. No apparatus should be in front of the fire building.

22.   **Firefighters should avoid a high-risk situation**
- a.   altogether.
- b.   when only a low-gain can be achieved.
- c.   rarely, since firefighting is always risky.
- d.   only b and c.
- e.   all of the above.

23.   **Most 30-minute self-contained breathing apparatus air cylinders actually have approximately how many minutes of usable air?**
- a.   15.
- b.   20.
- c.   25.
- d.   30.
- e.   None of the above.

24.   **For an accountability system to be effective it needs to know**
- a.   the number of personnel operating at the scene.
- b.   their approximate location.
- c.   the task or function they are performing.
- d.   only b and c.
- e.   all of the above.

25.   **A personnel accountability report**
- a.   should be taken routinely.
- b.   ensures accuracy of where personnel are operating.
- c.   will be checked against charts to ensure that everyone is accounted for.
- d.   only a and b.
- e.   all of the above.

26.   **A rapid intervention crew**
- a.   is a standby crew that is assigned for rescue of firefighters.
- b.   is intended to permit an immediate response to a firefighter in distress.
- c.   requires a full company of three or more firefighters.
- d.   only a and b.
- e.   all of the above.

27.   **Safe fireground operations demand**
- a.   working together as a team.
- b.   crew and company integrity at all times.
- c.   that units should enter and leave a fire area together or in teams of two or more.
- d.   only a and b.
- e.   all of the above.

28. **When a Mayday is sounded an Incident Commander should**

a. find out the reason for the sounding of the Mayday.

b. dedicate the current radio channel to the rescue efforts.

c. perform a personnel accountability report.

d. only a and c.

e. all of the above.

29. **Common cellars are where**

a. adjoining properties use a common cellar area with no fire walls between the properties.

b. separation may be accomplished by using wooden partitions to separate the individual cellar areas.

c. poor housekeeping by one storeowner can jeopardize all the areas above the cellars.

d. only a and c.

e. all of the above.

30. **A basement with a metal door flush with the pavement and locked from the inside can best be accessed by**

a. cutting through the door with a power saw using a metal cutting blade.

b. breaking out the concrete in each corner where the door is anchored and lifting or sliding the door aside.

c. using an explosive charge.

d. only a and b.

e. all of the above.

31. **At a cellar fire it may be necessary to place a line outside an exterior cellar door that is being used for ventilation since**

a. it can enter that doorway for fire extinguishment.

b. it can be used to play a stream into the doorway.

c. it can be used to prevent fire extension up the outside of the building.

d. only a and b.

e. all of the above.

32. **Cellar pipes or distributors can be used**

a. if unable or impractical to enter a cellar.

b. by feeling the floor for the hottest spot for the location of the fire below.

c. but must always be accompanied by a backup hose-line.

d. only a and c.

e. all of the above.

33.   **In regard to ventilation at a cellar fire:**
- a. Deadlights in the sidewalk can be broken out to effect ventilation.
- b. Breaking out the material under showcase windows in stores will assist in ventilation of a cellar.
- c. Ventilation in fighting a cellar fire can be effected by opening upper floor windows.
- d. Only a and b.
- e. All of the above.

34.   **Characteristics of garden apartments include**
- a. frame buildings of platform construction.
- b. brick veneer is a common wall covering.
- c. that they are often built in sections.
- d. only a and b.
- e. all of the above.

35.   **A roof overhanging the sidewalls of a building creates**
- a. eaves.
- b. bearing walls.
- c. nonbearing walls.
- d. only a and b.
- e. all of the above.

36.   **A poke-through**
- a. is an opening in fire-stopping material.
- b. negates compartmentation.
- c. can create a large non-fire-stopped area within a concealed space.
- d. only a and c.
- e. all of the above.

37.   **Problems associated with fires in garden apartments include**
- a. occupants often attempt to extinguish a fire themselves.
- b. that people assume that the fire department already has been called when a fire occurs.
- c. there may be interconnected void spaces throughout the building.
- d. only a and b.
- e. all of the above.

38.   **A common location for fires to start in garden apartments is in the storage areas normally located on the lowest level. This occurs because**
- a. these rooms are often rather large.
- b. they often contain a variety of combustible materials.
- c. storage can include flammable liquids and a variety of household articles.
- d. only a and c.
- e. all of the above.

39. **Row houses and town houses in general**
a. were built to meet the demand for economical housing.
b. share common party walls.
c. can be ordinary or frame construction.
d. only a and b.
e. all of the above.

40. **Characteristics of row houses in general include**
a. the rear of the row of houses on one street backs up to the rear of those on the opposing street.
b. there may be narrow alleyways between the rear of the row houses.
c. there may be limited access to the rear.
d. only a and b.
e. all of the above.

41. **At a working fire involving row houses it is not unusual to have smoke pushing from the cornices of many properties. This could indicate**
a. the immediate need for ventilation.
b. a common cockloft exists between properties.
c. that the fire has entered the adjacent properties.
d. only b and c.
e. all of the above.

42. **Fires in buildings undergoing renovations can cause problems since**
a. standpipe systems may be inoperable.
b. good housekeeping is not usually a priority on construction sites.
c. automatic detection and extinguishing equipment may have been removed.
d. only a and c.
e. all of the above.

43. **Ventilation problems that can be found in renovated buildings include**
a. modifications to window openings can impede ventilation and delay extinguishment.
b. brick can be used to seal basement windows.
c. windows on upper floors can be covered on the interior with drywall leaving the window intact on the exterior.
d. only a and b.
e. all of the above.

44.   **At a hotel or motel fire the Incident Commander must determine:**
a. What is burning?
b. What are the smoke and fire conditions in the hallways?
c. Where and how will the fire most likely spread?
d. What additional problems can develop?
e. All of the above.

45.   **Vacant buildings are**
a. frequently abandoned by owners who don't want to be found.
b. in varying states of decay.
c. often stripped of any contents of value.
d. only b and c.
e. all of the above.

46.   **Vacant buildings can create a concern since**
a. party walls can break down and a fire entering the cockloft or attic can spread to adjacent properties.
b. stability of exterior walls doesn't guarantee the stability of the roof.
c. these buildings can become a receptacles for trash.
d. only a and c.
e. all of the above.

47.   **Access to wildland urban interface locations may be impeded due to:**
a. Poor access roads.
b. Steep grades.
c. Stream crossings.
d. Terrain problems.
e. All of the above.

48   **The greatest danger in relation to magnetic resonance imaging for patients, hospital staff and firefighters is**
a. the low temperatures when expelling helium.
b. the neurological effect when the magnetic resonance imaging is operating.
c. the tremendous power of the magnet.
d. the lack of radio communications.
e. all of the above.

49.   **Today's codes incorporate life-safety features, yet a problem exists in many churches since**
a. many churches were built before the codes were written.
b. repairs and renovations may be substandard.
c. churches have undergone changes over the years.
d. only a and c.
e. all of the above.

50. **Where doorways exist between churches and other buildings**
a. the possibility of fire starting in one building and communicating to the other must be considered.
b. doors should be fire-rated.
c. doors should be self-closing.
d. only a and b.
e. all of the above.

51. **A major cause of church fires is**
a. arson.
b. electrical.
c. defective heating plants.
d. only a and b.
e. all of the above.

52. **Fires in churches or houses of worship create problems for firefighters due to**
a. overcrowding at certain religious celebrations.
b. separate services for adults and children might complicate the search-and-rescue problem.
c. hose-lines must not be stretched through doorways that are being used by fleeing occupants.
d. only a and c.
e. all of the above.

53. **1)** **During an interior attack at a church fire the exterior of the building must be monitored for changing conditions.**
2) **The fire attack at a church fire may generate enough steam to extinguish fire that has extended into the hanging ceiling.**
a. Both statements are true.
b. Both statements are false.
c. Only statement number one is true.
d. Only statement number two is true.

54. **Fires in churches or houses of worship create problems for firefighters due to the fact that**
a. a defensive attack must consider exposed buildings.
b. collapse zones will have to be set up.
c. interconnected buildings will require monitoring for fire extension.
d. only a and c.
e. all of the above.

55.   **Fires in public assemblies create exit problems because**
- a.   exit doors may be found locked for security reasons.
- b.   a firefighter relying on an exit as a secondary means of egress may find it locked.
- c.   windows are always available as exits, but harder to use.
- d.   only a and b.
- e.   all of the above.

56.   **In regards to an incident involving an active shooter, general considerations could include:**
- a.   Restricting the use of sirens within two blocks of the reported active shooter location for all emergency responders.
- b.   Seal off the area and set up a safe perimeter to prevent the entry of unsuspecting civilians.
- c.   Create zones on the exterior similar to hazardous materials responses.
- d.   Decide upon radio procedures and have designated frequencies.
- e.   All of the above.

57.   **In regard to fighting fires in commercial buildings and warehouses:**
- a.   Structures higher than the original fire building should not be considered an exposure concern.
- b.   Rapidly deteriorating conditions will necessitate the withdrawal of firefighters.
- c.   Tremendous radiant heat will allow rapid fire spread.
- d.   Only b and c.
- e.   All of the above.

58.   **In regard to fighting fires in commercial buildings and warehouses, hose-lines operated from the windows of threatened exposures will**
- a.   be of little value in extinguishing the fire.
- b.   place water on the fire.
- c.   prevent the extension of fire through that window.
- d.   only b and c.
- e.   all of the above.

59.   **1)**   **Strip malls may be built with masonry walls that parapet through the roof and serve as fire-stops.**
       2)   **Strip malls may be built as large store areas that are subdivided into smaller stores.**
- a.   Both statements are true.
- b.   Both statements are false.
- c.   Only statement number one is true.
- d.   Only statement number two is true.

60.   **In regard to a strip mall fire:**

     a.   An attack on the fire is almost always initiated from the front of the store.

     b.   The roof team can notify the Incident Commander if a parapet wall exists between the structures.

     c.   If a fire wall exists, spread to the exposures will never occur.

     d.   Only a and b.

     e.   All of the above.

61.   **A strip mall fire that has progressed past the incipient stage**

     a.   demands that the roof be opened to draw the fire to the exterior.

     b.   can spread to the adjacent stores via common roof spaces.

     c.   must be attacked with hose streams of sufficient volume to contain the fire.

     d.   only b and c.

     e.   all of the above.

62.   **The front of the store in an enclosed mall uses an open web metal security gates which**

     a.   allows security to be visually maintained from the common areas of the mall.

     b.   allows a fire to extend from one store to another.

     c.   has no effect on security or firefighting.

     d.   only a and b.

     e.   all of the above.

63.   **Large-area or department stores in enclosed malls may contain specialty departments within the main store. These may be shoe or camera sections or jewelry areas. In heavy smoke conditions a customer or firefighter entering this area**

     a.   could find a safe haven at these locations.

     b.   could become disoriented and trapped.

     c.   would be safe to await rescue from these locations.

     d.   only a and c.

     e.   all of the above.

64.   **When evacuating civilians from enclosed malls**

     a.   parents may enter dangerous areas to find their children, even though the children may have already been removed to safety.

     b.   wheelchairs and people with walkers can complicate evacuation.

     c.   people pushing and shoving will add to the confusion.

     d.   only a and c.

     e.   all of the above.

65. **Ventilation factors at enclosed malls include**

    a.   rear doors, if provided, are difficult to open from the exterior.

    b.   service doors or truck receiving areas that can be opened.

    c.   roof ventilation may require opening skylights, scuttles or the roof itself.

    d.   the mall ventilation system, if reversible, can be used.

    e.   all of the above.

66. **Fires in lumberyards create problems for firefighters because**

    a.   railroad tracks can mean dead-end water mains.

    b.   lumber piled with wood strips allow a fire to attack a much greater surface area.

    c.   stacked trusses create natural air spaces for fire to attack.

    d.   only a and c.

    e.   all of the above.

67. **Fires in lumberyards past the incipient stage create problems for firefighters since**

    a.   there is an enormous potential to escalate to major proportions.

    b.   it will necessitate a defensive attack.

    c.   exposures on the leeward side will need immediate protection.

    d.   only a and c.

    e.   all of the above.

68. **Fires in lumberyards past the incipient stage create water supply problems for firefighters since**

    a.   large quantities of water will be required.

    b.   water will be needed to reduce radiant heat, knock down the spreading fire, and reduce flying brands.

    c.   knowledge of the water supply will be needed.

    d.   only a and b.

    e.   all of the above.

69. **The core area in a high-rise building is**

    a.   the location where the utilities, shaftways, and elevators reach up through the building.

    b.   located in the center, front, rear, or side of the building.

    c.   located in the first floor lobby only.

    d.   only a and b.

    e.   all of the above.

70.   **In regard to sprinkler and standpipe systems in high-rise buildings:**
   a.   An adequate and continuous water supply needs to be delivered to these systems.
   b.   The fire department should know the size and capacity of these systems.
   c.   The building's fire pumps may feed these systems directly, or water tanks may be used.
   d.   Only a and b.
   e.   All of the above.

71.   **Pressure reducing valves**
   a.   reduce the volume of water available.
   b.   reduce the water pressure available.
   c.   are always adjustable.
   d.   only a and b.
   e.   all of the above.

72.   **The types of stairs found in high-rise buildings can be**
   a.   return type.
   b.   scissors-type.
   c.   U-return.
   d.   only b and c.
   e.   all of the above.

73.   **There are two basic design concepts for horizontal floor separations in high-rise buildings. They are**
   a.   open and shut areas.
   b.   core and noncore areas.
   c.   compartmentation and open area.
   d.   only b and c.
   e.   all of the above.

74.   **The plenum area in a high-rise building**
   a.   is created by installing a suspended ceiling.
   b.   is used for wiring, ducts, and other utilities.
   c.   may contain combustible material that will add fuel to a fire.
   d.   only a and b.
   e.   all of the above.

75.   **In regard to elevator usage by firefighters in high-rise buildings under fire conditions:**
   a.   Some fire departments forbid the use of freight elevators.
   b.   Some fire departments dictate when an elevator may be used.
   c.   A common rule in many fire departments is that the elevator should not be used if the fire is located on the first seven floors.
   d.   Only a and b.
   e.   All of the above.

76. **In regard to elevator usage by firefighters in high-rise buildings under fire conditions:**

    a.   Portable radio transmissions can affect electronic controls on some elevators.

    b.   If any doubt about the safe use of the elevator exists, climb the stairs.

    c.   Elevators always should be used so rescues can be performed in a timely manner.

    d.   Only a and b.

    e.   All of the above.

77. **Fire that has taken control of a floor area in excess of _____square feet is beyond the control of hand-held hose-lines.**

    a.   5,000.

    b.   7,500.

    c.   10,000.

    d.   12,500.

    e.   none of the above.

78. **When fighting high-rise fires the use of 45- or 60-minute self-contained air cylinders can affect firefighters since**

    a.   it can have a physically draining effect on firefighters.

    b.   firefighters base their stamina on the number of depleted air cylinders they normally use.

    c.   the duration of the air cylinders is not a factor for firefighters to worry about.

    d.   only a and b.

    e.   all of the above.

79. **High-rise building fires present unique problems that require implementation of specific functions in a command system. These could include:**

    a.   Lobby Control.

    b.   Stairwell Support.

    c.   Elevator Control.

    d.   Only a and c.

    e.   All of the above.

80. **Responsibilities of Lobby Control include**

    a.   control all building entry and exit points.

    b.   control and operate all elevator cars.

    c.   direct building occupants to proper ground-level safe areas.

    d.   only a and b.

    e.   all of the above.

81. **Hazardous materials incidents**
    a. can expose firefighters to uncontrolled situations.
    b. under fire conditions can be totally unpredictable.
    c. can quickly become a major problem if not handled expeditiously.
    d. only b and c.
    e. all of the above.

82. **The NFPA 704 marking system denotes the color blue as:**
    a. A fire hazard.
    b. A health hazard.
    c. An instability hazard.
    d. As a special designation.
    e. None of the above.

83. **The Chemical Transportation Emergency Center (CHEMTREC) can be called if a vehicle carrying hazardous materials is involved in an emergency. CHEMTREC, after being provided the name of a chemical, can assist the emergency responders by**
    a. providing the nature of the product.
    b. recommending steps to be taken in handling the early stages of an incident.
    c. contacting the shipper of the product for more detailed information.
    d. only a and c.
    e. all of the above.

84. **The Incident Commander must assess the most likely method of how hazardous materials can spread. This must consider**
    a. that high temperatures can cause some products to vaporize readily.
    b. runoff through sewers.
    c. water runoff as a result of firefighting efforts.
    d. only b and c.
    e. all of the above.

85. **A staging area designated at a hazardous materials incident must consider**
    a. topography.
    b. wind.
    c. accessibility.
    d. only a and b.
    e. all of the above.

86.   **At a hazardous materials incident**
- a.   a decontamination plan should be in place prior to anyone entering an area where exposure may occur.
- b.   medical personnel must protect themselves to prevent their being contaminated when treating someone who has been exposed.
- c.   prior to treatment and transporting to a hospital, decontamination of a patient must occur.
- d.   only b and c.
- e.   all of the above.

87.   **The weapons of a terrorist include armed attack. The acronym B-NICE stands for**
- a.   bombs, nuclear, inciting, chaos, and explosions.
- b.   biological, nuclear, incendiary, chemical, and explosive.
- c.   biological, nuclear, inciting, chemistry, explosions.
- d.   biological, nuisance, incendiary, chemical, and explosive.
- e.   none of the above.

88.   **Chemical weapons used by terrorists include**
- a.   sarin.
- b.   mustard gases.
- c.   cyanides.
- d.   only a and c.
- e.   all of the above.

89.   **Dispatch centers should develop a target hazard analysis that identifies a list of locations or various types of occupancies that could trigger a warning of a terrorist attack. This could include events occurring in**
- a.   historic buildings.
- b.   governmental targets.
- c.   controversial occupancies.
- d.   only a and b.
- e.   all of the above.

90.   **The earlier the recognition of the possibility of a terrorist event by firefighters the faster safeguards can be initiated. These safeguards should include**
- a.   setting up a hose-line for immediate decontamination of firefighters.
- b.   being aware of the direction of any water runoff that is occurring.
- c.   minimizing the number of personnel in the suspected areas.
- d.   only a and c.
- e.   all of the above.

91. **Unified Command will be needed for a successful conclusion to a terrorist event. Without Unified Command**
   a. there could be turf battles.
   b. evidence may be destroyed.
   c. duplication of effort by responders may occur.
   d. only a and c.
   e. all of the above.

92. **A community must be prepared to meet the threat of terrorism through stages of readiness. Stage 1 would best be described as**
   a. immediate response stage.
   b. warning stage.
   c. alert stage.
   d. recovery operations.
   e. none of the above

93. **A critique is meant to reconstruct events at an incident and assess how the fire department performed. The areas that should be reviewed include**
   a. what worked well.
   b. areas where improvement is needed.
   c. whether operational guidelines need revision.
   d. only a and b.
   e. all of the above.

94. **The informal critique can be held**
   a. at the incident scene.
   b. in the firehouse.
   c. at any convenient location.
   d. only b and c.
   e. all of the above.

95. **Information given at a formal critique can include**
   a. the information that was received by dispatch.
   b. what equipment was dispatched.
   c. the observations of the first-in company.
   d. only b and c.
   e. all of the above.

96. **Supervisors must recognize exceptional effort and praise those who have earned recognition since**
   a. firefighters want meaningful feedback.
   b. firefighters desire praise if it is earned.
   c. firefighters seek direction for self-improvement.
   d. only a and b.
   e. all of the above.

97. **The final report on a formal critique would address implementation of an Incident Management System in the following areas**
    a. staff positions created.
    b. groups created.
    c. divisions created.
    d. only a and b.
    e. all of the above.

98. **The final report on a formal critique would be concerned about apparatus and equipment in the following areas:**
    a. Was the apparatus properly placed and used?
    b. Could special equipment have completed the assignments in a safer manner?
    c. Did the equipment on hand meet the needs of the incident?
    d. Only a and b.
    e. All of the above.

99. **The "Lessons Learned" section of a formal critique should be written**
    a. so that it can become a learning process.
    b. for it to be beneficial to everyone.
    c. in a positive way.
    d. only a and b.
    e. all of the above.

100. **A large number of firefighter injuries are due to stress that occurs**
    a. when firefighters are in the firehouse.
    b. when critical decisions have to be made in a short period of time.
    c. en route to the incident scene.
    d. only a and c.
    e. all of the above.

ANSWER KEY Study Guide 5 for Chapters 6 through 11

Question	Answer	Page Reference	Question	Answer	Page Reference
1	A	210	26	D	253
2	C	210	27	E	255
3	C	215	28	E	257
4	A	218	29	E	263
5	E	219	30	B	264
6	D	220	31	C	264
7	E	222	32	E	265
8	E	223 – 224	33	D	266
9	E	226	34	E	268
10	E	229	35	A	269
11	E	230	36	E	269
12	E	231	37	E	269 – 270
13	B	231 – 232	38	E	271
14	E	232	39	E	274
15	E	233	40	E	275
16	E	235	41	E	278 – 279
17	D	237	42	E	282 – 283
18	E	240	43	E	284
19	C	243	44	E	292
20	D	243	45	E	297
21	C	248	46	E	297 – 298
22	B	250	47	E	305
23	A	251	48	C	322
24	E	251	49	E	346
25	E	252	50	E	347

Question	Answer	Page Reference	Question	Answer	Page Reference
51	E	350	76	D	437
52	E	351	77	C	441
53	A	352	78	D	443
54	E	355	79	E	445
55	D	369	80	E	499
56	E	381 – 382	81	E	458 – 459
57	D	399	82	B	463
58	D	400	83	E	464
59	A	404	84	E	466
60	D	404	85	E	469
61	E	405 – 406	86	E	470
62	D	411	87	B	485
63	B	412	88	E	485
64	E	412	89	E	486
65	E	413	90	E	486
66	E	423	91	E	488
67	E	424	92	C	490
68	E	425	93	E	527
69	D	430	94	E	527
70	E	431	95	E	528
71	D	431	96	E	529
72	E	432	97	E	530
73	C	435	98	E	531
74	E	436	99	E	531
75	E	436 – 437	100	B	532

1.   **Most firefighters are injured in collapses that**
   a.   involve large and newsworthy events.
   b.   involve total building collapse.
   c.   are caused by small and less spectacular events.
   d.   both a and b.
   e.   all of the above.

2.   1)   **The collapse zone must be recognized as a safety zone.**
     2)   **Once a collapse zone has been established, the area should be cordoned off. With the exception of chief officers no one else should enter this zone.**
   a.   Both statements are true.
   b.   Both statements are false.
   c.   Only statement number one is true.
   d.   Only statement number two is true.

3.   **Two or more floors fully involved in fire in a commercial and industrial building is a collapse indicator because**
   a.   exposed structural members are being attacked by fire.
   b.   there may be interconnection between floors for a conveyor belt.
   c.   large open areas favor a fast-moving fire.
   d.   only a and c.
   e.   all of the above.

4.   **Unprotected steel columns and beams exposed to heavy fire and high heat conditions can precipitate an early collapse because**
   a.   the heat of the fire causes them to expand.
   b.   steel will absorb the heat until its failure temperature is reached.
   c.   steel can push against an exterior wall causing it to lean outward.
   d.   only a and b.
   e.   all of the above.

5.   **The pressure of a fire can force or push smoke through any available openings, causing it to show through walls. The concern this causes firefighters is**
   a.   that collapse is imminent.
   b.   the movement and pressure of the smoke may indicate the possibility of a backdraft explosion.
   c.   that a strictly defensive attack should be employed.
   d.   only a and c.
   e.   all of the above.

6.   **Hose streams directed at a masonry wall**
- a.   can cause a breakdown in the wall.
- b.   can wash out mortar that is holding bricks in place.
- c.   can reduce its load-carrying capacity.
- d.   only a and b.
- e.   all of the above.

7.   **Previous fire damage in a building**
- a.   is cumulative in sapping a building's strength.
- b.   creates a dangerous situation for firefighters attacking a fire.
- c.   may be covered over and remain in walls of occupied buildings.
- d.   only b and c.
- e.   all of the above.

8.   **A firefighter fighting a fire in a building and hearing loud cracking noises realizes that this could indicate**
- a.   the fire is still burning.
- b.   that fire companies are still responding.
- c.   that parts of the building are failing.
- d.   only a and c.
- e.   all of the above.

9.   **When operating in an offensive mode, a buildup of water within a building requires**
- a.   that immediate action be taken to alleviate these conditions.
- b.   controlling the excess flow from hose-lines.
- c.   moving fire debris that is restricting runoff.
- d.   only b and c.
- e.   all of the above.

10.   **An eccentric load would include all of the below except:**
- a.   Wall signs.
- b.   Marquees.
- c.   Large ornate cornices.
- d.   Large flush mounted picture windows.
- e.   Corbelled brick.

11.   **An eccentric load acts like a downward thrust on a wall. It can be placed on a wall by**
- a.   a wall sign.
- b.   a marquee.
- c.   a roof sign.
- d.   only a and b.
- e.   all of the above.

12.   **Who should establish a rescue plan if a search is needed in a collapsed building?**
- a.   The first-arriving truck company officer.
- b.   The Incident Commander.
- c.   The Safety Officer.
- d.   Only a and b.
- e.   All of the above.

13. **When firefighters respond to a collapsed building with the possibility of people trapped they must**

    a.   check surrounding structures.

    b.   evacuate other structures if necessary.

    c.   check if other sections of the building are in danger of collapsing.

    d.   only a and b.

    e.   all of the above.

14. **Heavy rescue units have the necessary tools and equipment required for a complex rescue operation. If a heavy rescue unit is not available, consider using**

    a.   local utility companies.

    b.   truck companies that carry extrication equipment.

    c.   equipment under contract to the fire department.

    d.   only a and b.

    e.   all of the above.

15. **When confronted with a collapsed building and the possibility of people trapped, firefighters should**

    a.   ensure that after the removal of surface victims that voids are checked for victims.

    b.   ensure that searchers do not give up too early on the hope of finding survivors.

    c.   use thermal imaging cameras to assist in locating victims.

    d.   only a and c.

    e.   all of the above.

16. **At a building collapse incident**

    a.   a larger area can be searched if firefighters can connect the voids that occur in the collapse.

    b.   search and rescue teams must work in pairs.

    c.   too many search and rescue teams can cause coordination problems.

    d.   only b and c.

    e.   all of the above.

17. **The Incident Commander should consider appointing a Safety Officer**

    a.   on incidents that pose a significant hazard.

    b.   on incidents that could adversely affect the well-being of those operating at the scene.

    c.   whenever the Incident Commander does not want the responsibility for safety.

    d.   only a and b.

    e.   all of the above.

18. **A fire department Safety Officer should have a working knowledge of**
    a. building construction.
    b. strategy and tactics.
    c. human behavior.
    d. fire science.
    e. all of the above.

19. 1) **Asbestos is a carcinogen and was used for years in many facets of building construction.**
    2) **In a solid state asbestos presents little hazard, but as a dust it can be inhaled or ingested.**
    a. Both statements are true.
    b. Both statements are false.
    c. Only statement number one is true.
    d. Only statement number two is true.

20. 1) **Managing risk for emergency responders is something that is performed once and guidelines are written.**
    2) **Managing risk for emergency responders is a continuous process. Guidelines must be reviewed to ensure that they continue to meet the safety needs of the department.**
    a. Both statements are true.
    b. Both statements are false.
    c. Only statement number one is true.
    d. Only statement number two is true.

21. **The Safety Officer can enhance the monitoring at an incident scene by**
    a. waiting until a fire is under control before arriving on the scene.
    b. not entering the fire area until the fire is placed under control.
    c. using an incident safety check-off sheet and note pad.
    d. only a and b.
    e. all of the above.

22. **Safe operation of a ladder pipe dictates that firefighters should be located**
    a. at the head of the main ladder.
    b. on the ground operating ropes to control the ladder pipe.
    c. halfway up the main ladder to get a better view of the fire area.
    d. only a and c.
    e. all of the above.

23. **A building fire involving trapped civilians would be classified as a**
    a. low-gain situation.
    b. medium-gain situation.
    c. high-gain situation.
    d. only a and b.
    e. all of the above.

24. 1) **Accountability ensures control at an incident. Firefighters should operate in teams and not alone. Firefighters assigned to a truck company are the only firefighters permitted to freelance.**

    2) **Units must be part of a known plan and company officers must know where their firefighters are operating at all times.**

    a. Both statements are true.
    b. Both statements are false.
    c. Only statement number one is true.
    d. Only statement number two is true.

25. **Firefighters assigned to a rapid intervention crew need training in**

    a. using a thermal imaging device.
    b. using buddy breathing.
    c. basic extrication procedures.
    d. only b and c.
    e. all of the above.

26. **When firefighters are missing or lost, the overall focus of the incident scene must be centered on the missing firefighter/s. To assist in this operation**

    a. a personnel accountability system may be used to narrow down the location of the missing firefighter(s).
    b. the rapid intervention crew should be dispatched to the suspected location of the missing firefighter(s).
    c. someone should be placed in charge of the rescue effort.
    d. only a and b.
    e. all of the above.

27. 1) **If unprotected steel is supporting the first floor, a well-involved cellar fire can be fought with an interior attack since the steel will not present a threat of an early failure.**

    2) **Lightweight building components supporting a floor in residential and commercial properties will be a serious concern due to the potential of an early floor collapse.**

    a. Both statements are true.
    b. Both statements are false.
    c. Only statement number one is true.
    d. Only statement number two is true.

28. **A cellar fire may be recognized by**

    a. heavy smoke showing only on the upper floors of a building.
    b. the presence of heat and smoke at the first-floor level and the absence of visible fire.
    c. smoke emitting from the baseboards on lower floors and banking down on the top floor.
    d. only b and c.
    e. all of the above.

29.   **Before firefighters can remove self-contained breathing apparatus**

a.   the fire must be completely extinguished.
b.   visibility must be 100 percent.
c.   safe carbon monoxide levels must be approved through testing.
d.   only b and c.
e.   all of the above.

30.   **A distributor is**

a.   basically a large sprinkler head that is attached to a hose-line.
b.   lowered through a floor opening to combat a cellar fire.
c.   the perfect tool that will extinguish any cellar fire.
d.   only a and b.
e.   all of the above.

31.   **The term garden apartments is a reference that infers**

a.   that there are gardens surrounding the apartment buildings.
b.   that the building is set back from the roadway.
c.   that these types of buildings were originally found along the beer gardens in Germany.
d.   only a and b.
e.   all of the above.

32.   **Characteristics of garden apartments**

a.   roof assemblies can be pitched or flat roofs.
b.   interior partition walls are usually wood studs.
c.   walls between the adjoining sections can be of masonry construction.
d.   only a and b.
e.   all of the above.

33.   **Roof eaves**

a.   protect the walls below from the weather.
b.   can have openings that permit air circulation in the roof space.
c.   always have fire stops built in to prevent fire spread.
d.   only a and b.
e.   all of the above.

34.   **In regard to access to garden apartments:**

a.   The terrain can hamper access to the building.
b.   Illegal parking can prevent apparatus from entering parking areas.
c.   Apparatus may not be able to reach the sides or rear of some buildings.
d.   Only a and c.
e.   All of the above.

35. **When fighting fires in garden apartments the Incident Commander must determine**

   a.   the location and extent of the fire.

   b.   if the fire has entered floor or roof assemblies.

   c.   whether the fire has spread to adjacent apartments.

   d.   only a and c.

   e.   all of the above.

36. **A common location for fires to start in garden apartments is in the storage areas, which are normally located on the lowest level. Problems associated with the storage areas include that**

   a.   they usually consist of a wooden framework with wood planking and/or chicken wire.

   b.   gas and electric meters for the building can be found at this location.

   c.   facilities for washing and drying clothes can be located in the same room.

   d.   only a and b.

   e.   all of the above.

37. **In general, some newer style row houses and typically all town houses**

   a.   have garages located at the basement or first floor level.

   b.   that has garages in the rear have rear driveways for access.

   c.   have no access to the basement area from the rear of the property.

   d.   only a and b.

   e.   all of the above.

38. **Fire lapping out porch-front windows of row houses, which have a continuous wooden porch roof extending the length of the block**

   a.   can quickly spread to many properties.

   b.   demands a coordinated attack be made on arrival of the initial companies.

   c.   demands that hose-lines be deployed to stop the lateral spread.

   d.   only a and b.

   e.   all of the above.

39. **Building renovations can increase the fire load since**

   a.   vinyl wall coverings give off large amounts of deadly smoke when burning.

   b.   decorative paneling, once ignited, burns fiercely and can resemble a natural gas fire.

   c.   large quantities of wood framing are often used to convert mill buildings to apartments.

   d.   only a and b.

   e.   all of the above.

40.
1) A hotel fire will require that ventilation be initiated early to prevent mushrooming of fire and smoke on the top floors.

2) Since heavy smoke conditions will impact on the firefighting and evacuation efforts at a hotel fire, ventilation must be used to clear the smoke from the hallways and stairs.

a. Both statements are true.
b. Both statements are false.
c. Only statement number one is true.
d. Only statement number two is true.

41. **Vacant buildings can create a concern for**

a. the possibility of arson for profit in declining areas.
b. an invitation for trespassers to vandalize these buildings.
c. the reason that they can be death traps for firefighters.
d. only a and c.
e. all of the above.

42. **Vacant buildings can create a concern since**

a. they can be in a very weakened state.
b. the weight of firefighters and their operations may be too much for the building to bear.
c. the potential for an early collapse must be anticipated.
d. only a and c.
e. all of the above.

43. **Critical factors for wildland urban interface firefighting include:**

a. Weather.
b. Topography.
c. Fuel supply.
d. None of the above.
e. All of the above.

44.
1) Most surgical fires in operating rooms involve electrosurgery and initiate when the high energy electrosurgical unit or laser is accidentally activated.

2) Operating room fires may not generate sufficient heat to activate the sprinkler systems within the operating rooms.

a. Both statements are true.
b. Both statements are false.
c. Only statement number one is true.
d. Only statement number two is true.

45. **Immediate considerations at fires in nursing homes and assisted living facilities include:**

a. Ventilation.
b. Rescue/evacuation.
c. Fire containment.
d. Medical treatment.
e. All of the above.

46. **Gothic-style churches can**
a.   be one-story buildings equal to four stories or more in height.
b.   contain a hanging ceiling.
c.   have the pew area ceilings extend over 50-feet above them.
d.   none of the above.
e.   all of the above.

47. **Older wood frame churches often were built using**
a.   platform construction.
b.   balloon frame construction.
c.   log construction.
d.   only a and b.
e.   all of the above.

48. **Access to the church building may be a problem since**
a.   the church may be set back from the street.
b.   obstructions may restrict apparatus access.
c.   parked cars can hinder apparatus placement.
d.   only a and b.
e.   all of the above.

49. **Fires in churches or houses of worship create problems for firefighters due to the fact that in addition to worship services these buildings also are used as**
a.   day care centers.
b.   meeting places for senior citizens.
c.   food kitchens to feed the homeless.
d.   only a and b.
e.   all of the above.

50. **Church fires can be affected by a pyrolytic effect of wood. This refers to**
a.   old churches catching fire due to their age.
b.   wooden beams being changed to charcoal.
c.   an arsonist who torches churches.
d.   only a and c.
e.   none of the above.

51. **Fires in churches or houses of worship create problems for firefighters because**
a.   decorations for religious celebrations can add to the problem.
b.   a fire that is burning in concealed spaces is hard to locate.
c.   it is difficult to determine how extensive the fire is.
d.   only a and b.
e.   all of the above.

52. **Fires in churches or houses of worship create problems for firefighters because**
    a. a fast-moving interior fire will be difficult to control.
    b. hand-held hose-lines will be ineffective on fires past the incipient stage.
    c. light smoke inside the church could indicate the need for an exterior attack.
    d. only a and b.
    e. all of the above.

53. **Fires in churches or houses of worship create ventilation problems for firefighters because**
    a. steep-sloped roofs are too dangerous to operate on.
    b. roof venting should be performed by firefighters operating from a platform or a main ladder.
    c. firefighters may have to break stained-glass windows.
    d. only a and b.
    e. all of the above.

54. **Fires in public assemblies create life safety problems due to the fact that**
    a. combustible interior furnishings will allow a fire to spread rapidly.
    b. a building's heating and air-conditioning systems can spread fire and smoke.
    c. individuals confronted with fire conditions will react in the way they feel will best safeguard themselves.
    d. only b and c.
    e. all of the above.

55. **In regards to an incident involving an active shooter, general considerations could include:**
    a. Restricting the use of sirens within two blocks of the reported incident for all emergency responders.
    b. Sealing off the area and set up a safe perimeter to prevent the entry of unsuspecting civilians.
    c. Creating zones on the exterior similar to hazardous materials responses.
    d. Deciding upon radio procedures and have designated frequencies.
    e. All of the above

56. **In regard to fighting fires in commercial buildings and warehouses:**
    a. Fire control can often be achieved through aggressive firefighting methods.
    b. Fire control can be achieved when assisted by good building construction.
    c. Large, wide-open areas allow the spread of fire.
    d. Only b and c.
    e. All of the above.

57. **A deteriorating condition in a commercial building or warehouse fire that could signal a switch to a defensive operation would be:**
    a. A large body of fire that is not diminishing with attacking hose-lines.
    b. Water from the hose-lines turning to steam with no obvious effect on the fire.
    c. Continued or worsening high heat and/or heavy smoke conditions.
    d. None of the above
    e. All of the above.

58. **In regard to fighting fires in commercial buildings and warehouses:**
    a. High-piled stock will restrict the penetration of exterior streams.
    b. The success of a defensive attack depends on the number of windows through which the fire can be fought.
    c. The success of a defensive attack depends upon the size of the structure itself.
    d. Only a and b.
    e. All of the above.

59. **The term taxpayer refers to**
    a. a strip mall.
    b. a strip store.
    c. an arcade.
    d. only a and b.
    e. all of the above.

60. **Strip malls may be built with canopies or large decorative false fronts with signs attached to the front of the buildings. It has been found that these attachments**
    a. are often constructed with lumber.
    b. can collapse under heavy fire conditions.
    c. are usually fire-stopped
    d. only a and b.
    e. all of the above.

61. **If there is a report of civilians trapped in a fire in the rear of a store in a strip mall it will cause problems for the firefighters since**
    a. the rear doors are usually padlocked.
    b. the initial attack on the fire and the rescue of those trapped inside must occur through the rear of the store.
    c. it may by difficult finding the correct store location in the rear area.
    d. only a and b.
    e. all of the above.

62. **The enclosed shopping mall causes firefighters problems because**

    a.   it creates a severe life hazard in case of fire.

    b.   many people are unfamiliar with the mall and its exits.

    c.   these malls are always located in remote areas with a minimal water supply.

    d.   only a and b.

    e.   all of the above.

63. **Enclosed shopping malls have a high turnover of businesses, which prompts constant renovations. These renovations can lead to**

    a.   the presence of construction material that will add to the fire load.

    b.   shutting down the sprinkler system for specific periods.

    c.   numerous arson fires.

    d.   only a and b.

    e.   all of the above.

64. **Some malls are designed with separate stockrooms. If no stock area is provided, there is a tendency to**

    a.   overload display areas with stock.

    b.   have congestion in the aisles.

    c.   use shelving to cordon off an area for storage.

    d.   only a and b.

    e.   all of the above.

65. **The resource needs for complete evacuation of an enclosed mall can quickly overtax the responding firefighters. Life safety can be assisted by**

    a.   the use of police and security to assist in the evacuation.

    b.   an aggressive attack on the fire.

    c.   bringing hose-lines into the interior mall area to contain the fire and protect the mall walkways.

    d.   only b and c.

    e.   all of the above.

66. **Fires in supermarkets create problems for firefighters due to**

    a.   the potential for injuries to civilians and firefighters.

    b.   disabled employees and patrons may need assistance.

    c.   some supermarkets operate continuously.

    d.   only a and b.

    e.   all of the above.

67. **Fires in lumberyards past the incipient stage create problems for firefighters since**

    a.   a fast-moving fire must be fought from the flanks.

    b.   wind is unpredictable and can change direction without warning.

    c.   it may not be possible to maintain a position in front of a fire.

    d.   only b and c.

    e.   all of the above.

68.   **For a fire department to be able to control a fire past the incipient stages in a fully enclosed lumberyard requires**

    a.   a big city fire department.

    b.   a properly installed and operating sprinkler system.

    c.   large-diameter hose-line.

    d.   only a and b.

    e.   all of the above.

69.   **A core constructed high-rise building could consist of a**

    a.   center core.

    b.   side core.

    c.   rear core.

    d.   only b and c.

    e.   all of the above.

70.   **In regard to sprinkler and standpipe systems in high-rise buildings:**

    a.   A dry-standpipe system may be provided in fire towers.

    b.   It should be noted on the preplan whether the dry standpipes are interconnected.

    c.   The water tanks supplying these systems may have a dedicated water supply.

    d.   Only b and c.

    e.   All of the above.

71.   **In a high-rise building what unit/s is responsible for checking that the standpipe outlets are closed above the fire floor?**

    a.   Engine company.

    b.   Truck company.

    c.   Heavy rescue company.

    d.   Haz mat company.

    e.   All of the above.

72.   **In regard to stair shafts in high-rise buildings**

    a.   scissors-type stairs consist of two sets of stairs in a common stair shaft.

    b.   scissors stairs may alternate floors with each set of stairs in the stair shaft.

    c.   a sufficient amount of stairs are located throughout the building to enable the total evacuation of a high-rise building.

    d.   only a and b.

    e.   all of the above.

73. **In addition to portable radios, the fire department can use additional communications systems to assist them when fighting a high-rise fire. This can include**

    a.   cellular telephones.

    b.   hard-wire systems.

    c.   elevator intercoms.

    d.   only a and c.

    e.   all of the above.

74. **It could be correctly stated that the open-space concept used in today's modern high-rise office building means that**

    a.   each floor is basically wide open.

    b.   partitions may be used to separate the floor area.

    c.   this design allows fire to spread rapidly throughout the entire floor.

    d.   only b and c.

    e.   all of the above.

75. **In regard to elevator usage by firefighters in high-rise buildings under fire conditions:**

    a.   A common rule in many fire departments is that, if a fire is located above the seventh floor, the use of the elevators is at the discretion of the Incident Commander.

    b.   If an elevator is used, it must have firefighter's service.

    c.   A common rule in many fire departments is that the elevator should not be used if the fire is located on the first seven floors.

    d.   Only a and b.

    e.   All of the above.

76. **The term auto-exposure refers to**

    a.   fire extending to adjacent buildings.

    b.   fire extending horizontally to involve attached areas.

    c.   fire extending via windows to the floor above.

    d.   only b and c.

    e.   all of the above.

77. **Lightweight deluge guns that are set up to attack high-rise fires that are past the control of handheld hose-lines will**

    a.   provide a greater water flow to absorb the heat of the fire.

    b.   create water supply problems.

    c.   be restricted by movable partitions set between workstations.

    d.   only a and c.

    e.   all of the above.

78. **When considering life safety of occupants during a high-rise fire the Incident Commander must decide**

    a.   whether to evacuate the occupants.

    b.   which floors to evacuate.

    c.   whether the occupants should be protected in place.

    d.   only a and b.

    e.   all of the above.

79. 1) **A working high-rise fire demands the immediate implementation of an Operations Section Chief.**
    2) **The staging area in a high-rise building should be a minimum of two floors below the fire floor.**
    a. Both statements are true.
    b. Both statements are false.
    c. Only statement number one is true.
    d. Only statement number two is true.

80. **Stairwell Support is**
    a. used to move equipment from the lobby to the staging area via stairwells.
    b. accomplished by stationing a firefighter every two floors.
    c. a very laborious and tiring assignment.
    d. only b and c.
    e. all of the above.

81. **Concerns at hazardous materials incidents are that**
    a. contact with these substances by firefighters is often difficult to detect.
    b. firefighters may be confronted with a wide range of hazards.
    c. there is the potential for serious complications because of the presence of unknown chemicals and their interaction with other products.
    d. only b and c.
    e. all of the above.

82. **A resource to assist firefighters in identifying hazardous materials is provided by the DOT in its *Emergency Response Guidebook.* In regard to this publication**
    a. a copy should be carried on every emergency response apparatus.
    b. it gives vital initial information on placarding systems.
    c. it assists in identifying products within a facility through the identification of placard numbers.
    d. only a and b.
    e. all of the above.

83. **Many responses to hazardous materials incidents occur in laboratories within hospitals, research facilities, or industrial plants. These facilities**
    a. usually have no recirculation of air within the building.
    b. have ventilation systems which exchanges the air within these laboratories every few minutes.
    c. have relatively small amounts of a product that are used in experiments.
    d. only a and c.
    e. all of the above.

84. **The hot zone at a hazardous materials incident**
  a. is the immediate danger area.
  b. is an exclusion area and should be considered contaminated.
  c. will require those entering this area to have a high level of personal protection.
  d. only a and c.
  e. all of the above.

85. **Firefighters' personal protective equipment**
  a. has limitations when dealing with hazardous materials.
  b. can be saturated or penetrated by hazardous materials.
  c. allows toxic chemicals to cause chronic exposure to a firefighter if the gear is not decontaminated.
  d. only a and c.
  e. all of the above.

86. **There is a distinct difference between a hazardous materials incident and a terrorist incident. In the terrorist incident**
  a. the release of hazardous materials has been done intentionally.
  b. there is a strong possibility of the presence of secondary devices.
  c. there is always notification to the authorities of the incident in advance of its occurrence by the terrorists.
  d. only a and b.
  e. all of the above.

87. **Biological weapons of terrorism include**
  a. anthrax.
  b. the plague.
  c. deadly viruses.
  d. only a and b.
  e. all of the above.

88. **Chemical weapons used by terrorists include**
  a. arsenic.
  b. chlorine.
  c. phosgene.
  d. only a and c.
  e. all of the above.

89. **Dispatch centers should develop a target hazard analysis list that includes locations or types of occupancies that could trigger a warning of an impending terrorist attack. This could include events occurring**
  a. in airports.
  b. aboard airplanes.
  c. in places of assembly.
  d. only a and b.
  e. all of the above.

90. **Without advance information, if a terrorist act involves a biological weapon, it is highly likely that**
   a.   initial responders will become part of the problem.
   b.   first-responder units on a medical call can be severely exposed.
   c.   exposure may involve delayed symptoms.
   d.   only a and b.
   e.   all of the above.

91. **Incident scene safety and scene security are a must. If an armed terrorist attack has occurred, the police can assure the firefighters scene safety by which of the following?**
   a.   Having the terrorists in custody.
   b.   Confirming that the terrorists have left the scene.
   c.   Having the bomb squad check the scene thoroughly for secondary devices.
   d.   Only b and c.
   e.   All of the above.

92. **What radioactive element have experts identified as the most likely to be used in a dirty bomb?**
   a.   Uranium.
   b.   Cesium 137.
   c.   Plutonium.
   d.   Only a and c.
   e.   None of the above.

93. **Cues that can assist in identifying a methamphetamine lab include:**
   a.   Strong odors like cat urine or ammonia.
   b.   Unusual amounts of clear glass containers being brought into the home.
   c.   Windows blackened out or covered by aluminum foil, plywood, sheets, blankets, etc.
   d.   Secretive/protective area surrounding the residence.
   e.   All of the above.

94. **A formal critique should be held**
   a.   after most major emergencies.
   b.   after most significant events.
   c.   as soon after the incident as possible.
   d.   only b and c.
   e.   all of the above.

95. **Participants at a formal critique who commanded units or were responsible for a major segment of the operation should discuss**
   a.   their observations.
   b.   problems they encountered.
   c.   orders they received or had given.
   d.   only b and c.
   e.   all of the above.

96.   **A written report sharing a critique's findings should be shared with**
   a.   the members who attended the critique.
   b.   other fire department members.
   c.   mutual aid departments who responded.
   d.   only a and c.
   e.   all of the above.

97.   **A final report of a formal critique would be concerned about strategy and tactics in the following areas:**
   a.   Strategies developed.
   b.   Tactics initiated.
   c.   What problems arose that required special attention.
   d.   Only a and b.
   e.   All of the above.

98.   **A final report of a formal critique would be concerned about resources in the following areas:**
   a.   Were requests for additional resources timely?
   b.   Was there a time when certain functions could not be performed due to a lack of resources at the scene?
   c.   Was a insufficient amount of resources called for?
   d.   Only a and c.
   e.   All of the above.

99.   **A committee should routinely compare the formal critique reports to previous ones**
   a.   to see if there are recurring problems.
   b.   to see if patterns are developing which need to be addressed.
   c.   for disciplinary purposes.
   d.   only a and b.
   e.   all of the above.

100.   **Firefighters associate fireground events to their own situations. When a fellow firefighter is injured or killed at an incident**
   a.   they often question why the injury occurred to someone else and not them.
   b.   they may falsely blame themselves for causing the injury.
   c.   a sense of guilt adds to their stress.
   d.   only a and b.
   e.   all of the above.

ANSWER KEY Study Guide 6 for Chapters 6 through 11

Question	Answer	Page Reference	Question	Answer	Page Reference
1	C	210	26	E	255
2	C	211	27	D	263
3	E	215 – 216	28	D	264
4	E	216	29	C	265
5	B	219	30	D	265
6	E	220	31	B	268
7	E	220 – 222	32	E	268 – 269
8	C	223	33	D	269
9	E	225	34	E	269
10	D	226 – 227	35	E	270
11	D	226 – 227	36	E	271
12	B	229	37	D	275
13	E	230	38	E	278
14	E	231	39	E	283
15	E	232	40	A	293
16	E	233	41	E	297
17	D	234 – 235	42	E	298
18	E	235	43	E	306
19	A	237	44	A	326
20	D	242 – 243	45	E	333
21	C	244	46	E	346
22	B	250	47	B	347
23	C	250	48	E	348
24	D	252	49	E	348
25	E	253 – 254	50	B	350

ANSWER KEY Study Guide 6 for Chapters 6 through 11

Question	Answer	Page Reference	Question	Answer	Page Reference
51	E	351	76	C	440
52	D	352	77	E	441
53	E	354	78	E	444 – 445
54	E	367 – 368	79	A	447 – 448
55	E	381 – 382	80	E	451
56	E	397	81	E	459
57	E	398	82	E	464
58	E	400	83	E	464 – 465
59	D	403	84	E	466 – 467
60	D	404 – 405	85	E	468
61	E	405	86	D	484
62	D	409	87	E	485
63	D	412	88	E	485
64	E	412	89	E	486
65	E	412 – 413	90	E	486 – 487
66	E	417	91	E	488
67	E	424	92	B	495
68	B	426	93	E	503
69	E	430	94	E	527
70	E	431	95	E	528
71	B	432	96	E	530
72	D	432 – 433	97	E	531
73	E	434	98	E	531
74	E	435	99	D	531
75	E	436 – 437	100	E	532

Study Guide 7 Chapters 6 through 11

1.   **When dealing with buildings that have a high potential for collapse**
   a.   speed is essential, and firefighters must use the fastest methods available to enter and fight fires in these buildings.
   b.   aggressiveness will be counterproductive.
   c.   if any doubt about firefighters commencing an interior attack exists, then do it quickly before things change.
   d.   only a and c.
   e.   all of the above.

2.   **In regard to a building's strength and the potential for building collapse:**
   a.   Bearing walls are the strongest part.
   b.   The corners of a building is its strongest point.
   c.   Non-bearing walls are the strongest point.
   d.   Only a and c.
   e.   None of the above.

3.   **High heat and heavy smoke conditions coupled with inadequate ventilation indicate a collapse potential since**
   a.   the fire cannot be found.
   b.   visibility is limited.
   c.   these conditions set the stage for a backdraft.
   d.   only a and c.
   e.   all of the above.

4.   **When a steel beam reaches its failure temperature**
   a.   it will sag and can slide down the inside of the bearing wall it previously rested upon.
   b.   it can exert a tremendous pressure and push the bearing wall on which it previously rested violently outward.
   c.   it can cause parts of the wall it previously rested upon to be projected past a 100-percent collapse zone as it collapses.
   d.   only a and b.
   e.   all of the above.

5.   **When a new wall crack is discovered it indicates**
   a.   that the building has been under stress for a long period of time.
   b.   that the building is experiencing movement or failure at that time.
   c.   that there should be no concern, one wall crack is nothing to worry about.
   d.   only a and b.
   e.   all of the above.

6.   **A wall spreader is installed in a masonry wall**
- a. when the building is over two stories in height.
- b. when the wall is first constructed, as additional support for the wall.
- c. when a wall has developed structural problems, so the spreaders can act as stabilizers for the wall.
- d. only b and c.
- e. all of the above.

7.   **Observation of windows and doors prior to entering a building**
- a. can indicate their location for finding a secondary means of egress.
- b. can indicate whether they are out of plumb, indicating a structural problem.
- c. is only necessary for members of the truck company.
- d. only a and b.
- e. all of the above.

8.   **Plaster sliding off of walls would indicate**
- a. a condition that requires more monitoring before any actions are taken.
- b. a severe collapse condition.
- c. an immediate withdrawal from the building of all personnel.
- d. only b and c.
- e. all of the above.

9.   **Water can be retained in a building through absorption. This can mean that**
- a. the building will absorb some water.
- b. the stock within a building can absorb large quantities of water.
- c. little serious thought should be given to this factor. This is overrated and should not be considered a problem for the IC.
- d. only a and b.
- e. all of the above.

10.   **An impact load can occur by**
- a. a main ladder being dropped upon a cornice.
- b. a firefighter chopping with an axe to open a roof.
- c. a firefighter jumping onto a roof surface.
- d. only b and c.
- e. all of the above.

11. **The best way to minimize and handle problems involving a collapse rescue is**
   a.   by having firefighters in various areas of the collapse site and having them decide the best method of extricating those trapped.
   b.   by using an Incident Management System.
   c.   by having multiple Safety Officers and letting them decide individually the best way to handle the situation.
   d.   only b and c.
   e.   all of the above.

12. **When confronted with a collapsed building and the possible need for rescues within the building the Incident Commander should ensure**
   a.   that the fire department takes control of the utilities supplying the collapsed structure when possible.
   b.   that minimal amounts of water be used to control a fire.
   c.   that utility personnel enter and secure the various utilities as firefighters make rescues.
   d.   only a and b.
   e.   all of the above.

13. **The various types of floor collapse are**
   a.   pancake, V-type, lean-to, unsupported floor.
   b.   pancake, total, lean-to, outward.
   c.   pancake, unsupported, V-type, outward.
   d.   V-type, lean-to, inward-outward, pancake.
   e.   none of the above.

14. **Firefighters operating at collapsed buildings must be concerned about safety. Common injuries at these types of incidents are caused by**
   a.   operating in areas deficient in oxygen.
   b.   puncture wounds.
   c.   secondary collapse.
   d.   only a and b.
   e.   all of the above.

15. **The delegation of safety at an incident scene must occur**
   a.   when the complexities of an incident could prevent the Incident Commander from handling safety on a priority basis.
   b.   to ensure continuity of this important task.
   c.   to allow the Incident Commander time for other important decisions.
   d.   only a and b.
   e.   all of the above.

16. **Unsafe acts being performed by firefighters at an incident scene**
   a. must be assessed to see how dangerous they are.
   b. must be evaluated to see if the task can still be completed.
   c. must be corrected immediately.
   d. only a and b.
   e. all of the above.

17. **1)** **Risk can be totally eliminated at an incident scene.**
   2) **A role of the incident scene Safety Officer is to observe operations and act as a risk manager.**
   a. Both statements are true.
   b. Both statements are false.
   c. Only statement number one is true.
   d. Only statement number two is true.

18. **The constant surveillance of an incident scene by visual observation is referred to as**
   a. 360-degree walk-around.
   b. monitoring.
   c. checking it out.
   d. only a and b.
   e. all of the above.

19. **Tunnel vision can be defined as**
   a. being blinded when entering a tunnel on a sunny day.
   b. becoming so engrossed in a particular phase of an operation that one fails to see the big picture.
   c. not constantly looking over your shoulder to keep abreast of what is occurring behind you.
   d. only b and c.
   e. all of the above.

20. **When evaluating risk it should be weighed against**
   a. the gain to be achieved.
   b. the benefit to be achieved.
   c. how long an operation will take.
   d. only a and b.
   e. all of the above.

21. **An accountability system for firefighters operating on an incident scene must be in place to track personnel. It will assist in**
   a. knowing what everyone is doing.
   b. knowing where everyone is operating.
   c. notification of those endangered.
   d. only a and b.
   e. all of the above.

22.   **The acronym PAR stands for**

a.   preparing a rescue.
b.   personnel accountability report.
c.   preparation, allocation, rescue.
d.   personnel allowance reserves.
e.   none of the above.

23.   **Firefighters assigned to rapid intervention crews should note**

a.   all means of entry and egress from the building.
b.   the location of raised ground and aerial ladders.
c.   the location of windows that access roof areas.
d.   only a and b.
e.   all of the above.

24.   **Factors or cues observed by the Incident Commander that could indicate the need for upgrading the RIC from a single company to a task force are:**

a.   A working fire in a commercial building.
b.   A working fire in a high-rise building.
c.   A working fire in a large residential structure.
d.   A significant or unusual fire situation.
e.   All the above.

25.   1)   **The cellar of a building can be one large open area, which can be more conducive to firefighting.**

2)   **When attacking a cellar fire the reach of a hose stream will be reduced if the height of storage reaches near the ceiling.**

a.   Both statements are true.
b.   Both statements are false.
c.   Only statement number one is true.
d.   Only statement number two is true.

26.   **In regard to high expansion foam being used to combat cellar fires:**

a.   It has had some success in extinguishing cellar fires.
b.   Ventilation must be provided in front of the foam for proper distribution.
c.   Hose-lines must be shut down to prevent breaking down the foam.
d.   Only a and b.
e.   All of the above.

27.   **Garden apartments have features similar to**

a.   town houses.
b.   row houses.
c.   single family homes.
d.   only a and b.
e.   all of the above.

28. **Characteristics of garden apartment's roof assemblies include**
    a. solid wood beams.
    b. wooden "I" beams.
    c. steel bar-joist.
    d. only a and b.
    e. all of the above.

29. **Roof eaves can be covered with**
    a. plywood.
    b. aluminum soffit material.
    c. vinyl soffit material.
    d. only a and b.
    e. all of the above.

30. **In regard to garden apartment construction:**
    a. The first floor may be partially below ground level.
    b. The building's terrain may be landscaped in the rear.
    c. Compartmentation will protect the occupants on the upper floors.
    d. Only a and b.
    e. All of the above.

31. **A common location for fires to start in garden apartments is in the storage areas normally located on the lowest level. Problems associated with the storage areas include**
    a. housekeeping is often lax.
    b. the door to this room is often propped open.
    c. a fire originating in this area can spread upward through ceiling openings.
    d. only b and c.
    e. all of the above.

32. **Characteristics of row houses and town houses in general include**
    a. they contain a common party wall.
    b. that the front and rear walls are normally nonbearing walls.
    c. that the party walls are normally nonbearing walls.
    d. only a and b.
    e. all of the above.

33. **Characteristics of row houses in general include that**
    a. stairs are often steep.
    b. hallways can be long and narrow.
    c. they contain tight quarters that can restrict firefighting operations.
    d. only a and b.
    e. all of the above.

34.   **When roof operations are used for fighting fires in row houses and town houses firefighters must consider that**

     a.   if adjacent buildings are the same height, roof access can be gained from adjoining buildings.

     b.   adjoining roofs allow firefighters a secondary means of egress.

     c.   roof access always must be achieved directly onto the fire building via aerial ladders.

     d.   only a and b.

     e.   all of the above.

35.   **Temporary shoring**

     a.   can be used to support a structure until a permanent correction can be accomplished.

     b.   too often these stopgap measures become permanent.

     c.   fire originating around the shoring can cause a fast and deadly collapse.

     d.   only a and c.

     e.   all of the above.

36.   **Typically, hotel and motel fires**

     a.   are contained to one guest room.

     b.   will involve more than one guest room.

     c.   will be past the incipient stages prior to the arrival of the fire department.

     d.   all of the above.

     e.   none of the above.

37.   **Vacant buildings create a concern for firefighters due to the fact that**

     a.   further damage to these buildings can be caused by the weather.

     b.   rain water attacking structural members can weaken the structure.

     c.   there is the potential to affect attached properties structurally.

     d.   only a and c.

     e.   all of the above.

38.   **From a risk versus gain perspective at a wildland urban interface fire, buildings can be placed into which of the following categories:**

     a.   Structures that are not threatened.

     b.   Structures that are threatened but defensible and have the potential of being saved.

     c.   Structures that are threatened but not defensible and too dangerous to protect.

     d.   Structures that are owned by the fire department.

     e.   Only a, b and c.

39.   **For firefighters to be successful at a fire in a nursing home or an assisted living facility requires:**
   - a. Preplanning and fire prevention education.
   - b. Staff training and properly operating building protective systems.
   - c. Early fire department notification.
   - d. Implementation of proper strategic and tactical considerations.
   - e. All of the above.

40.   **How many levels of trauma centers are there in the United States?**
   - a. One.
   - b. Two.
   - c. Three.
   - d. Four.
   - e. Five.

41.   **Inspections of churches should include**
   - a. climbing into hanging ceilings to see what hazards may be contained there.
   - b. activating alarm systems.
   - c. checking if renovations meet applicable codes.
   - d. only b and c.
   - e. all of the above.

42.   **The fire officer, along with a representative of the church, should decide what valuables must be protected or removed should a fire occur. This must include placing a value on the stained-glass windows since**
   - a. stained-glass windows may be worth a fortune.
   - b. prior thought must be given whether to break or try to protect these windows.
   - c. firefighters like to break windows.
   - d. only a and b.
   - e. all of the above.

43.   **The church may be only one building in a complex surrounded by other buildings. There can be interconnections between the church and the rectory, parsonage, convent, or school. They may be linked together by**
   - a. doorways.
   - b. tunnels.
   - c. above ground walkways.
   - d. only a and c.
   - e. all of the above.

44.   **The leading contributing factor of major fires in churches is**
- a.   the large area on the interior of the building.
- b.   delayed detection or delayed notification of the fire department.
- c.   poor firefighting.
- d.   only a and b.
- e.   all of the above.

45.   **Fires in churches or houses of worship create problems for firefighters since**
- a.   containing the fire in the early stages will be necessary for an interior attack to be successful.
- b.   the fire department may experience problems gaining entry to the church.
- c.   it can be difficult to ascertain what is burning.
- d.   only b and c.
- e.   all of the above.

46.   **The ceiling in a gothic-style church is referred to as a hanging ceiling because it**
- a.   is suspended from the timber truss.
- b.   depends on the truss for its support.
- c.   can easily be entered by firefighters to extinguish a fire.
- d.   only a and b.
- e.   all of the above.

47.   **Fires in churches or houses of worship create problems for firefighters due to the fact that**
- a.   a concern of the congregation is the valuable artifacts in the church and they may enter the building to rescue these items.
- b.   firefighters should attempt to remove chalices, torahs, and other relics.
- c.   a firefighter's only duty is to extinguish the fire.
- d.   only a and b.
- e.   all of the above.

48.   **The penal institution preplan should have contingencies for a variety of situations that could occur at a penal institution. These include:**
- a.   Riots.
- b.   Fires.
- c.   Medical emergencies, including mass casualty incidents.
- d.   Power outages.
- e.   All of the above.

49.   **Fires in public assemblies create problems for firefighters due to**

    a.   the potential for overcrowding in these assemblies.
    b.   exit doors congested with people.
    c.   the maximum occupancy loads for these buildings are often ignored.
    d.   only a and c.
    e.   all of the above.

50.   **Fires in public assemblies create exit problems because**

    a.   conference rooms may have only a single entrance and exit point.
    b.   exit doors are always unlocked and easy to find.
    c.   exits may be blocked by tables, curtains, or large planters.
    d.   only a and c.
    e.   all of the above.

51.   **Fires in public assemblies create problems for firefighters because**

    a.   overcrowding will compound evacuation problems.
    b.   the lighting may be dimmed for an aesthetic effect.
    c.   coordination is needed to assure that firefighting efforts do not endanger occupants.
    d.   only a and c.
    e.   all of the above.

52.   **The operation of a sprinkler system when fighting fires in commercial buildings and warehouses can have an affect on the fire causing**

    a.   immediate extinguishment.
    b.   fire control.
    c.   a fire that is hampered by the sprinkler system, but still spreads at floor level.
    d.   only a and b.
    e.   all of the above.

53.   **In regard to fighting fires in commercial buildings and warehouses that contain fire doors:**

    a.   They often are blocked open.
    b.   Vertically sliding or overhead rolling doors can close automatically when activated by heat.
    c.   Automatic fire doors closing behind firefighters may disorient them.
    d.   Only a and c.
    e.   All of the above.

54.   **In regard to fighting fires in commercial buildings and warehouses:**
- a.   Success of a defensive attack depends upon knowing the effective reach of a hose stream.
- b.   Interior partitions will restrict the penetration of exterior streams.
- c.   Buildings up to 250 feet deep can be reached easily with all master stream devices.
- d.   Only a and b.
- e.   All of the above.

55.   **A strip mall**
- a.   contains a row of stores.
- b.   has a parking lot directly in front of the stores.
- c.   common to find no sprinkler protection.
- d.   only a and c.
- e.   all of the above.

56.   **In regard to a strip mall fire**
- a.   a fast-moving fire can quickly involve more than one store.
- b.   if rescue is not a problem, and the fire has gained control of an exposed building, it will be necessary to consider that store lost.
- c.   a blitz attack may be the difference between success and failure.
- d.   only a and b.
- e.   all of the above.

57.   **1)**   **Common cocklofts and attics should be anticipated at a fire in a strip mall. A fire entering a non-fire-stopped space above the ceiling can spread horizontally.**

      2)   **Using a trench cut can sometimes control a fire involving a large non-fire-stopped cockloft in a strip mall. This tactic can be advantageous and can be accomplished with a minimal number of resources.**
- a.   Both statements are true.
- b.   Both statements are false.
- c.   Only statement number one is true.
- d.   Only statement number two is true.

58.   **The enclosed shopping mall causes firefighters problems due to**
- a.   stores include restaurants, auto parts, linen, or paint stores.
- b.   there are no in-rack sprinkler systems.
- c.   the fire load will require an upgraded response from the fire department.
- d.   only a and b.
- e.   all of the above.

59. **Downtown enclosed malls usually have large, multistoried parking garages adjacent to or beneath them. These parking garages can create problems for firefighters because**
   a. they usually are built to accommodate cars only, and restrict entry of fire apparatus.
   b. firefighters responding for a vehicle fire or an accident with resultant injuries may be delayed trying to find the best location to gain entry.
   c. a buildup of carbon monoxide can develop in below grade garages if ventilation fans fail.
   d. only a and b.
   e. all of the above.

60. **Firefighters responding to a reported fire in an enclosed mall can find that**
   a. parking lots are virtually jammed with cars.
   b. the use of service entrances can provide ready access.
   c. hordes of people are attempting to leave the mall as firefighters try to enter.
   d. only a and c.
   e. all of the above.

61. **When evacuating civilians from enclosed malls**
   a. ramps can be used for ease of movement of people.
   b. one firefighter can safely lead 20 or more ambulatory people to safety.
   c. it may be necessary to assign rescuers on a one-to-one basis with the handicapped.
   d. only b and c.
   e. all of the above.

62. **Water supply for fires in enclosed malls in suburban or rural areas will need to consider the possibility of**
   a. long hose-line stretches.
   b. pump relay or water tender operations.
   c. using large-diameter hose-line.
   d. only a and c.
   e. all of the above.

63. **Problems associated with supermarket fires are:**
   a. Storage may be placed in the store proper causing congestion.
   b. Trailers containing storage may remain at loading docks restricting access.
   c. Access to the rear of the supermarket may be limited.
   d. Only a and b.
   e. All of the above.

64. **Firefighters' actions at a fire in a lumberyard in a few piles of lumber should include**

    a.   an aggressive attack on the fire.

    b.   a 2½-inch hose-line should be the minimum size hose-line used.

    c.   a blitz attack with a master stream can be most effective.

    d.   only a and b.

    e.   all of the above.

65. **Fires in lumberyards past the incipient stage create problems for firefighters since**

    a.   radiant heat can ignite structures a distance away.

    b.   shifting winds can affect exposures on more than one front.

    c.   convected hot gases can take fiery brands a distance from the fire starting other fires.

    d.   only a and b.

    e.   all of the above.

66. **Fires in high-rise buildings built today**

    a.   generate much higher temperatures than older high-rise buildings.

    b.   the building materials do not absorb the heat of the fire.

    c.   burn the same as those built in the early part of the 20th century.

    d.   only a and b.

    e.   all of the above.

67. **The best way to protect life and control or extinguish high-rise fires is**

    a.   through an aggressive attack with hose-lines.

    b.   through the installation and maintenance of an automatic sprinkler system.

    c.   by having personnel to constantly monitor building systems for alarms.

    d.   only b and c.

    e.   all of the above.

68. **Standpipe systems in high-rise buildings may contain pressure-reducing valves which**

    a.   reduce the water pressure.

    b.   reduce the volume of water.

    c.   may be adjustable.

    d.   only a and c.

    e.   all of the above.

69. **Stairwell pressurization can be accomplished by outside air introduced at**

    a.   the bottom of the stair shaft.

    b.   the top of the stair shaft.

    c.   various levels throughout the stair shaft.

    d.   only a and b.

    e.   all of the above.

70. **High-rise buildings constructed today use computers to control**
    a. fire-protection systems.
    b. security systems.
    c. other building systems.
    d. only a and c.
    e. all of the above.

71. **The heating, ventilation, and air-conditioning system used in high-rise buildings**
    a. uses outside filtered air brought into the building.
    b. recirculates air from occupied areas.
    c. uses the plenum area for recirculation of air.
    d. only b and c.
    e. all of the above.

72. **In regard to elevator usage by firefighters in high-rise buildings under fire conditions:**
    a. When using elevators members must have their self-contained breathing apparatus ready for immediate use.
    b. When ascending in an elevator, it should be stopped at random floors to check that the controls are operating properly.
    c. Firefighters should exit the elevator at least two floors below the fire floor and then climb the stairs to the fire floor.
    d. Only a and c.
    e. All of the above.

73. **When fighting a high-rise fire, firefighters should**
    a. attach their own hose-line to the standpipe on the floor below the fire.
    b. be prepared to encounter high heat and heavy smoke.
    c. attach hose-lines stretched to backup the initial hose-line to the standpipe on the fire floor.
    d. only a and b.
    e. all of the above.

74. **A high-rise fire can extend to the floor above through**
    a. spaces between the concrete floor and the exterior curtain wall.
    b. auto-extension via the windows.
    c. poke-throughs in the concrete floors.
    d. only b and c.
    e. all of the above.

75. 1) **A fire on a lower floor of a high-rise building that is past the point of control by an interior attack and poses a threat of auto-extension of fire to the floor above might require an exterior attack.**

2) **An exterior attack on a fire in a high-rise building is never an option. The building systems must be used and firefighters must attack the fire from within.**

    a. Both statements are true.
    b. Both statements are false.
    c. Only statement number one is true.
    d. Only statement number two is true.

76. **In regard to a search for occupants at a high-rise building fire:**

    a. Searches consume time and energy.
    b. A list of building occupants is helpful.
    c. Forcible entry may be required.
    d. Only a and b.
    e. All of the above.

77. **Success of a high-rise operation will be determined by the ability of the fire department to move resources and equipment up into the building in a timely manner. This is the responsibility of the**

    a. Incident Commander.
    b. Operations Section Chief.
    c. Logistics Section Chief.
    d. Safety Officer.
    e. None of the above.

78. **First-arriving units at a suspected hazardous materials incident**

    a. must define the problem.
    b. must be alert for signs of hazardous materials.
    c. should seek information from all available sources.
    d. only a and b.
    e. all of the above.

79. **A fixed facility can offer some benefits to initial responders. This includes**

    a. NFPA 704 identification numbers if marked on tanks.
    b. SDS sheets should be available on site.
    c. there may be fire protective systems in place.
    d. only b and c.
    e. all of the above.

James P. Smith

80. On suspected hazardous materials responses
a.   rain could spread a leaking material.
b.   responders should approach the site from an upwind location.
c.   when possible responders should use a downhill approach.
d.   only a and b.
e.   all of the above.

81. The warm zone at a hazardous materials incident
a.   allows an area for decontamination of personnel and equipment.
b.   is the immediate danger area.
c.   requires that access to the zone be controlled.
d.   only a and c.
e.   all of the above.

82. The term BLEVE refers to
a.   Big Levitation Expected Very Early.
b.   Boiling Liquid Explosion of Vapor Exponents.
c.   Boiling Liquid Expected Vapor Explosion.
d.   Boiling Liquid Expanding Vapor Explosion.
e.   None of the above.

83. Firefighters must document hazardous materials exposures. This data
a.   will be useful in determining specific exposures during an individual's career.
b.   may assist in diagnosing a later ailment.
c.   should list any chemicals that the firefighter has been exposed to.
d.   only a and b.
e.   all of the above.

84. At a hazardous materials incident the Incident Commander must determine what civilian dangers exist and
a.   the weather conditions.
b.   how long it will take to evacuate.
c.   whether it would be best to protect them in place rather than evacuate.
d.   only a and c.
e.   all of the above.

85. Problems associated with storage tanks at refineries and tank farms containing various types of oil are:
a.   Boilover, expansion, and frothover.
b.   Boilover, slopover, and expansion.
c.   Boilover, slopover, and frothover.
d.   Expansion, slopover, and frothover.
e.   None of the above.

86. 1) **The United States Department of Justice describes terrorism in part as "a violent act or an act dangerous to human life, in violation of criminal laws of the United States."**

    2) **Domestic terrorism is classified as those groups or individuals whose terrorist acts are directed at elements of our government or population with foreign direction.**

    a. Both statements are true.
    b. Both statements are false.
    c. Only statement number one is true.
    d. Only statement number two is true.

87. **Terrorists use incendiary weapons since they are economical and easily acquired. They can be thrown, or triggered remotely by**

    a. chemical means.
    b. electronic means.
    c. mechanical means.
    d. only b and c.
    e. all of the above.

88. **Terrorists can use explosives to disperse other agents, which include**

    a. biological.
    b. chemical.
    c. incendiary.
    d. only b and c.
    e. all of the above.

89. **The best methods of protection for firefighters at a chemical or biological terrorist event is**

    a. time.
    b. distance.
    c. shielding.
    d. only b and c.
    e. all of the above.

90. **The earlier the recognition of the specific agents and the concentration of the product used in a terrorist event, the faster**

    a. decontamination guidelines can be established.
    b. emergency medical operations for treatment of patients can be initiated.
    c. it will assist in the search for evidence.
    d. only a and b.
    e. all of the above.

91. **Civilians who are suspected of being contaminated at a terrorist event should**
    a. have their contaminated clothing placed in a bag, tagged, and have it secured to establish a chain of custody.
    b. realize that their clothing may contain evidence needed for a criminal investigation.
    c. be detained at the scene until decontaminated and medically cleared.
    d. only a and b.
    e. all of the above.

92. **A community must be prepared to meet the threat of terrorism through stages of readiness. Stage 3 would best be described as**
    a. immediate response stage.
    b. warning stage.
    c. alert stage.
    d. recovery operations.
    e. none of the above

93. **Clandestine Drug Laboratories may be categorized as an:**
    a. Active laboratory.
    b. Inactive lab.
    c. Abandoned lab.
    d. Only a and b.
    e. All of the above.

94. **Hurricanes are classified by categories, based on their wind speeds and potential to cause damage. How many categories are there?**
    a. 3.
    b. 4.
    c. 5.
    d. 6.
    e. 7.

95. **A recorder should take notes at a formal critique to prepare the final report. Special emphasis should be placed on**
    a. problems encountered.
    b. solutions.
    c. failed remedies.
    d. only a and b.
    e. all of the above.

96.   **The final report on a formal critique can be divided into three sections. The first part is a narrative account describing**
   a.   conditions.
   b.   problems encountered.
   c.   life safety considerations.
   d.   fire department actions.
   e.   all of the above.

97.   **A final report on a formal critique relating to medical assignments would be concerned with:**
   a.   What medical problems were dealt with?
   b.   Was this a mass casualty incident?
   c.   Was rehabbing of firefighters needed?
   d.   Only a and b.
   e.   All of the above.

98.   **A final report on a formal critique relating to outside agencies would be concerned with:**
   a.   What agencies were requested and responded?
   b.   Did the requested agencies meet the needs of the incident?
   c.   How can outside agencies better assist at an incident in the future?
   d.   Only a and b.
   e.   All of the above.

99.   **In general, firefighters are extremely dedicated individuals who**
   a.   hold themselves to a very high set of standards.
   b.   are action-oriented.
   c.   seek immediate results.
   d.   only a and b.
   e.   all of the above.

100.   **At a formal critical incident stress debriefing**
   a.   it must be strictly confidential.
   b.   no records should be kept.
   c.   only those firefighters directly involved in the incident and the debriefing team should attend.
   d.   only a and c.
   e.   all of the above.

ANSWER KEY Study Guide 7 for Chapters 6 through 11

Question	Answer	Page Reference	Question	Answer	Page Reference
1	B	210	26	E	266
2	B	215	27	D	268
3	C	216	28	E	269
4	E	216 – 217	29	E	269
5	B	219	30	D	269
6	D	220	31	E	271
7	D	222	32	D	275
8	D	223	33	E	276
9	D	225	34	D	279
10	E	228	35	E	283 – 284
11	B	229	36	A	292
12	D	231	37	E	298
13	A	233	38	E	308
14	E	234	39	E	330
15	E	235	40	D	338
16	C	235	41	E	346
17	D	242	42	D	347
18	B	243	43	E	347
19	B	247	44	B	348
20	D	250	45	E	351
21	E	251	46	D	352
22	B	252	47	D	354 – 355
23	E	254	48	E	361
24	E	255	49	E	367
25	A	263	50	D	369

ANSWER KEY Study Guide 7 for Chapters 6 through 11

Question	Answer	Page Reference	Question	Answer	Page Reference
51	E	370	76	E	445
52	E	398	77	C	448
53	E	399	78	E	459
54	D	400	79	E	464
55	E	403	80	E	465 – 466
56	E	405	81	D	467
57	C	406 - 407	82	D	467
58	E	409	83	E	468
59	E	410	84	E	469
60	E	412	85	C	478
61	E	412	86	C	484 – 485
62	E	413	87	E	485
63	E	417	88	E	485
64	E	424	89	E	486
65	E	424	90	E	487
66	D	430	91	E	489
67	B	430	92	A	491
68	E	431	93	E	503
69	E	433	94	C	511
70	E	435	95	E	528
71	E	435 – 436	96	E	530
72	E	437	97	E	531
73	E	439	98	E	531
74	E	440	99	E	532
75	C	441	100	E	533

Study Guide 8 Chapters 6 through 11
MID-TERM

1.   **The decision to enter and fight fires in structures prone to collapse where no life hazard exists**
   a.   is no different than any other structure fire.
   b.   is foolhardy.
   c.   is what firefighters do.
   d.   only a and c.
   e.   all of the above.

2.   1) **The 90-degree wall collapse assumes that a wall will fall outward its entire height encompassing a 90-degree angle.**
     2) **The inward-outward collapse has the top of the wall falling into the building and the lower part of the wall outward and away from the building.**
   a.   Both statements are true.
   b.   Both statements are false.
   c.   Only statement number one is true.
   d.   Only statement number two is true.

3.   **In regards to collapse zones:**
   a.   They can be called safety zones.
   b.   These zones should be cordoned off.
   c.   Experienced chief officers can enter these zones.
   d.   Only a and b.
   e.   All of the above.

4.   **A roof assembly commonly found in noncombustible buildings is**
   a.   lightweight wood truss.
   b.   steel bar joist truss.
   c.   timber truss.
   d.   only b and c.
   e.   all of the above.

5.   **As the intensity of the fire increases, flames may become visible through cracks in the exterior wall. This can indicate**
   a.   that the fire is in its final stage and will be burned out shortly.
   b.   that the fire is increasing in size and is attacking the structural members in the void spaces.
   c.   that hose-lines are reaching the seat of the fire.
   d.   only a and c.
   e.   all of the above.

6.   **A wall spreader being installed today in a wall with a structural problem in most jurisdictions requires**

a.   that the master builder on the scene make the decision on the proper installation.

b.   that the master brick layer on the scene make the decision on the proper installation.

c.   that an engineer design and submit drawings for approval.

d.   only a and c.

e.   all of the above.

7.   **After entering a building, firefighters may find interior doors stuck or jammed. This could indicate**

a.   that the occupants have left in a hurry.

b.   that the doors were of cheap quality.

c.   that the building has shifted.

d.   only a and b.

e.   all of the above.

8.   **Buildings under construction, renovation, or demolition**

a.   may have unprotected structural members.

b.   may have structural supports replaced with temporary shoring.

c.   can have varying stages of fire protection.

d.   only a and b.

e.   all of the above.

9.   **Absorbent bales contained within a fire building subjected to large amounts of water can**

a.   drastically increase in weight, placing extremely high loads on supporting members.

b.   expand and push out load-bearing walls or supporting columns.

c.   shift and fall when expanding bales are piled one upon another.

d.   only a and c.

e.   all of the above.

10.   **A building collapse can affect a fire by**

a.   resulting in near extinguishment of the fire.

b.   involving a much larger area and spreading the fire.

c.   allowing the extension of fire to nearby exposures.

d.   only b and c.

e.   all of the above.

11.   **The Safety Officer's responsibilities at a collapse search include**

a.   recognizing warning signs of collapse.

b.   recommending when firefighters should be rotated to avoid fatigue.

c.   monitoring members for signs of critical incident stress.

d.   only a and c.

e.   all of the above.

12. **When confronted with a collapsed building and the possibility that firefighters are trapped, the Incident Commander should**

a. have a personnel accountability report taken.
b. dedicate a radio channel to the rescue effort.
c. have personnel listen for the sound of a PASS device.
d. only a and c.
e. all of the above.

13. **When confronted with a collapsed building and a need to locate victims the Incident Commander should ensure**

a. that surface victims are first to be rescued.
b. that they restrict anyone from walking on the pile of debris since it can cause further injury to those trapped.
c. that those rescued are interviewed for additional information.
d. only b and c.
e. all of the above.

14. **At a building collapse incident**

a. vibrations from all sources must be eliminated.
b. hose-lines should be stretched as a precautionary measure.
c. noise will not be a concern.
d. only a and b.
e. all of the above.

15. **A fire department Safety Officer**

a. ensures that emergency scene safety is addressed.
b. reviews all injuries and determines their cause.
c. reviews all injuries looking for ways to prevent similar occurrences.
d. only b and c.
e. all of the above.

16. **Monitoring of the incident by the Safety Officer**

a. keeps firefighters from performing unsafe acts due to the Safety Officer's presence.
b. maintains a systematic safety analysis of the scene.
c. has little effect on safe operations.
d. only a and b.
e. all of the above.

17. **All aspects of an incident scene operation should be analyzed for risk. Research can include departmental records, the experience of fire department members, and**

a. national journals.
b. statistics from the United States Fire Administration.
c. statistics from the National Fire Protection Association.
d. only b and c.
e. all of the above.

18. **Monitoring of an incident scene by the Safety Officer includes**
    a. visual observation.
    b. listening to communications.
    c. using instrumentation to detect or identify problems.
    d. only a and c.
    e. all of the above.

19. **In regard to apparatus placement:**
    a. It must consider the use of the apparatus.
    b. An apparatus should not be placed where it will impede movement of other apparatus.
    c. A fire hydrant should not be used if it is directly in front of a burning building.
    d. Only a and b.
    e. All of the above.

20. **Risk to firefighters can be classified as**
    a. low.
    b. medium.
    c. high.
    d. only b and c.
    e. all of the above.

21. 1) **Risk can be avoided by nonintervening actions, which is a method of protecting firefighters.**
    2) **The classification of "risk versus gain" when used by responders always places everything into clear-cut categories.**
    a. Both statements are true.
    b. Both statements are false.
    c. Only statement number one is true.
    d. Only statement number two is true.

22. **An accountability system can be simple or complex, depending on the needs of the individual fire department. For a system to be effective**
    a. it must be a nationally recognized system.
    b. it must be practiced.
    c. all personnel must be familiar with it.
    d. only b and c.
    e. all of the above.

23. 1)   **OSHA, in regard to the two-in-two-out rule, states that "once firefighters begin the attack on an interior structural fire, the atmosphere is assumed to be immediately dangerous to life and health and the two-in-two-out rule applies."**

    2)   **The team outside of the immediately dangerous to life and health area must be located on the exterior of the fire building**

   a.   Both statements are true.
   b.   Both statements are false.
   c.   Only statement number one is true.
   d.   Only statement number two is true.

24.   **Interior cellar doorways**
   a.   in residential, multistory buildings, usually will be located beneath the stairs to the upper floors.
   b.   in one-story, residential properties often will be located in the kitchen.
   c.   in commercial properties are always located in the rear of the first floor.
   d.   only a and b.
   e.   all of the above.

25.   **When confronted with a cellar fire it should be noted that**
   a.   the best protection is afforded by sprinklers.
   b.   entrance to the basement area may be required to shut down the building's utilities.
   c.   if sprinklers are present the system should be pressurized.
   d.   only b and c.
   e.   all of the above.

26.   **In regard to ventilation at a cellar fire:**
   a.   It may be accomplished by opening an outside cellar door.
   b.   It may be accomplished by opening cellar windows.
   c.   It may be accomplished by cutting a hole in the floor beneath a window on the first floor.
   d.   Only a and b.
   e.   All of the above.

27.   **In regard to roof eaves on garden apartments:**
   a.   They can be in close proximity to the windows on the top floor.
   b.   Fire can easily extend from the top floor windows into the roof space via the eaves.
   c.   They may not contain fire-stopping.
   d.   Only a and b.
   e.   All of the above.

28. **In regard to draft-stopping in roof areas in garden apartments:**
- a. It can be used to compartmentalize an attic into smaller areas.
- b. It may be found depending upon when the building was built.
- c. It may be found depending upon the size of the roof area and the local code.
- d. Only a and b.
- e. All of the above.

29. **A problem when fighting fires in garden apartments can include**
- a. an immediate rescue effort by the first-arriving units could delay an attack on the fire.
- b. long hose-line stretches will be required if these buildings are set back from the street.
- c. a relatively problem-free and easy fire to handle.
- d. only a and b.
- e. all of the above.

30. **The initial strategies when fighting garden apartment fires**
- a. must be based upon the best method of protecting the occupants and firefighters.
- b. demands close coordination between suppression, search and rescue, and ventilation crews.
- c. should consider the possible spread of fire to common areas and other apartments.
- d. only a and b.
- e. all of the above.

31. **With regard to older row houses in general:**
- a. They were built with full-sized lumber.
- b. Their flooring is full-sized one-by-four-inch tongue and groove boards.
- c. Their roof planks are an inch thick.
- d. Only b and c.
- e. All of the above.

32. **In regard to row houses in general:**
- a. These structures may be four stories in height and equipped with standpipes.
- b. They are built on narrow streets that restrict entry of apparatus.
- c. Long stretches mean that preconnected hose-lines may not reach the fire.
- d. Only b and c.
- e. All of the above.

33.   **Front cornices of row houses**
- a.   can be constructed of sheet metal.
- b.   are often non-fire-stopped and may interconnect from one property to the next.
- c.   may have smoke pushing from the cornices of many properties at a working fire.
- d.   only b and c.
- e.   all of the above.

34.   **Light and air shafts contained in row houses and town houses**
- a.   present a danger to firefighters operating via windows.
- b.   have windows within these shafts that are useless for ventilation purposes.
- c.   may permit fire to extend via windows within the shaft to the next building.
- d.   only a and c.
- e.   all of the above.

35.   **Problems associated with buildings undergoing renovations without obtaining the proper permits include**
- a.   plans will not be reviewed.
- b.   workmanship can be substandard.
- c.   material may be used that is not in compliance with the applicable codes.
- d.   only a and c.
- e.   all of the above.

36.   **Problems associated with renovated buildings include**
- a.   sprinkler systems that are out of service can allow a fire to get past the incipient stage.
- b.   fire-stops may be removed.
- c.   the problems will be no different than any other building.
- d.   only a and b.
- e.   all of the above.

37.   **Runaway children or homeless people often occupy so-called vacant buildings. They often select structures**
- a.   that are in good condition.
- b.   that are in the poorest of conditions.
- c.   where the building owners are not around.
- d.   only b and c.
- e.   all of the above.

38.   **When fighting a fire in a vacant building, firefighters should**
- a. assume that the building is unoccupied.
- b. still undertake a primary search to ensure that the building is unoccupied.
- c. if occupied, make rescues in the safest manner possible realizing the hazards.
- d. only b and c.
- e. all of the above.

39.   **Hospitals are required to have an EOP. The abbreviation EOP stands for?**
- a. Emergency occupational partners.
- b. Emergency operating plans.
- c. Existing occupational practices.
- d. Engineered operating plans.
- e. None of the above.

40.   **A mass casualty incident is one in which:**
- a. The number of patients exceeds the level of stabilization and care that can normally be provided in a fast and timely manner.
- b. Responding units will not be sufficient to handle the many injured.
- c. Nearby hospitals will be overloaded.
- d. The number of patients involved will require significant implementation of emergency medical services (EMS) personnel in the affected community.
- e. All of the above.

41.   **Gothic-style churches often contain hanging ceilings that are**
- a. formed with plaster spread over wood or metal lath.
- b. 50 feet or higher above the nave or pew area.
- c. are the same as a dropped ceiling.
- d. only a and b.
- e. all of the above.

42.   **In regard to churches with votive candles**
- a. they produce an open flame that can be dangerous.
- b. many churches have changed to an electronic bulb-type candle.
- c. they have proven to be a harmless feature in a church.
- d. only a and b.
- e. all of the above.

43.   **Older wood frame churches were often built of**
- a. platform construction.
- b. balloon frame construction.
- c. log construction.
- d. only a and b.
- e. all of the above.

44. **The most common cause of church fires is**
- a. electrical.
- b. defective heating systems.
- c. arson.
- d. lightning.
- e. none of the above.

45. **Church fires can be affected by pyrolytic effect of wood. This refers to**
- a. old churches catching fire due to their age.
- b. wooden beams being changed to charcoal.
- c. an arsonist who torches churches.
- d. only a and c.
- e. none of the above.

46. **When fighting fires in churches or houses of worship, firefighters should consider**
- a. using thermal imaging to assist in locating the fire.
- b. stretching the first hose-line to the seat of the fire.
- c. stretching the second hose-line to directly above the fire area.
- d. only a and b.
- e. all of the above.

47. **1) During an interior attack at a church fire the exterior of the building must be monitored for changing conditions. The smoke emitting from the building can give telltale signs.**
2) The fire attack at a church fire may generate enough steam to extinguish fire that has extended into the hanging ceiling.
- a. Both statements are true.
- b. Both statements are false.
- c. Only statement number one is true.
- d. Only statement number two is true.

48. **The ceiling in a gothic-style church is referred to as a hanging ceiling. It**
- a. is suspended from the timber truss.
- b. depends on the truss for its support.
- c. can easily be entered by firefighters to extinguish a fire.
- d. only a and b.
- e. all of the above.

49. **Fires in public assemblies create problems for firefighters because**

a. combustible interior furnishings can permit a fire to spread rapidly.

b. nightclubs and restaurants can be located on the top floors of high-rise buildings.

c. maze-like configurations of some public assembly buildings can confuse people disoriented by smoke.

d. only a and c.

e. all of the above.

50. **Fires in public assemblies create exit problems due to**

a. conference rooms may have only single entrance and exit point.

b. exit doors are always unlocked and easy to find.

c. exits may be blocked by tables, curtains, or large planters.

d. only a and c.

e. all of the above.

51. **Should a fire force the dismissal of students for the day, officials should consider:**

a. The age of the children.

b. If parental notification will be required.

c. That high school students can usually be dismissed without parental notification.

d. That a thorough roll call is taken.

e. All of the above.

52 **In regards to school violence involving an active shooter, the fire department will generally:**

a. Assist the police since this is a police incident.

b. Become part of Unified Command if it is created.

c. Keep dispatch informed of conditions.

d. Assist in subduing the active shooter.

e. Only a, b and c.

53. **In regard to fighting fires in commercial buildings and warehouses:**

a. Success of a defensive attack depends upon knowing the effective reach of a hose stream.

b. Interior partitions will restrict the penetration of exterior streams.

c. Buildings up to 200-feet deep can be reached easily with all master stream devices.

d. Only a and b.

e. All of the above.

54.   **The term taxpayer refers to a (an)**
- a.   strip mall.
- b.   strip store.
- c.   arcade.
- d.   only a and b.
- e.   all of the above.

55. **1)** **Strip malls may be built with masonry walls that parapet through the roof and serve as a fire-stop.**
      2) **Strip malls may be built as large store areas that are subdivided into smaller stores.**
- a.   Both statements are true.
- b.   Both statements are false.
- c.   Only statement number one is true.
- d.   Only statement number two is true.

56.   **A strip mall may**
- a.   be built on a concrete slab.
- b.   contain a basement.
- c.   contain common basements.
- d.   only a and b.
- e.   all of the above.

57.   **At a strip mall fire an initial visible inspection through the front windows of the fire building and the adjoining properties can**
- a.   detect the presence of smoke or fire.
- b.   be considered a primary search.
- c.   give an indication of the severity and involvement of the fire.
- d.   only a and c.
- e.   all of the above.

58.   **A strip mall fire that has progressed past the incipient stage**
- a.   demands that the roof be opened to draw the fire to the exterior.
- b.   can spread to the adjacent stores via common roof spaces.
- c.   strip mall roof top operations can be a dangerous.
- d.   only b and c.
- e.   all of the above.

59. **1)** **Entry into the rear of a store in a strip mall can be challenging. Since theft and vandalism are a constant problem, rear doors are fortified to resist illegal entry.**
      2) **In strip malls of frame construction it is sometimes easier to gain rear entry through the rear wall alongside the door rather than through the door itself.**
- a.   Both statements are true.
- b.   Both statements are false.
- c.   Only statement number one is true.
- d.   Only statement number two is true.

60. **The aim of ventilation at a strip mall fire is to open the roof directly above the fire area to relieve pressure and vent the fire to the outside. Ventilating a roof that contains lightweight components is a dangerous task, and safety measures should be employed, which could include**
    a. working from the platform of a tower ladder.
    b. operating from an adjoining roof that is separated by a masonry fire wall.
    c. working quickly before the roof can collapse.
    d. only a and b.
    e. all of the above.

61. **The enclosed shopping mall can cause problems because**
    a. it creates a severe life hazard in case of fire.
    b. many people are unfamiliar with the mall and its exits.
    c. such malls are always located in remote areas with a minimal water supply.
    d. only a and b.
    e. all of the above.

62. **Large, sprawling malls need a system of identification for mall areas. A typical system can be based upon**
    a. interior color coding.
    b. exterior color coding.
    c. interior and exterior color coding.
    d. exterior numerical coding.
    e. interior and exterior alphabetical coding.

63. **Large-area or department stores in enclosed malls may contain specialty departments within the main store. These may be shoe or camera sections or jewelry areas. In heavy smoke conditions a customer or firefighter entering this area**
    a. could find a safe haven at these locations.
    b. could become disoriented and trapped.
    c. would be safe to await rescue there.
    d. only a and c.
    e. all of the above.

64. **Most search and rescue operations at supermarket fires are:**
    a. Initiated from the front of the store.
    b. Realize that store customers rarely observe the posted exit signs.
    c. Primary searches must include walk-in freezers.
    d. Only b and c.
    e. All of the above.

65. **Fires in lumberyards create problems for firefighters due to**

    a.   railways alongside a lumberyard fire must be shut down.

    b.   hose-lines may have to be stretched over railroad tracks.

    c.   if located on a waterway it will provide an ample supply of water.

    d.   only a and c.

    e.   all of the above.

66. **Fires in lumberyards create problems for firefighters due to**

    a.   the large quantity of highly combustible material in exposed exterior piles.

    b.   the lumber and material stored in combustible buildings.

    c.   fires develop quickly, with intense heat and rapid fire spread.

    d.   only a and c.

    e.   all of the above.

67. **Fires in lumberyards past the incipient stage will create problems for firefighters since**

    a.   urban lumberyards can be surrounded by properties that restrict access and firefighting capabilities.

    b.   brand patrols must be organized to check areas downwind for fire extension.

    c.   a fast-moving fire may require the relocation of apparatus.

    d.   only a and c.

    e.   all of the above.

68. **A core-constructed high-rise building built today can weigh less than**

    a.   5 pounds per square foot.

    b.   8 pounds per square foot.

    c.   14 pounds per square foot.

    d.   20 pounds per square foot.

    e.   25 pounds per square foot.

69. **The best method of extinguishing a high-rise fire is**

    a.   by an aggressive fire attack with 1 ¾-inch hose-lines

    b.   by an aggressive fire attack with 2 ½-inch hose-lines.

    c.   installation and maintenance of a sprinkler system.

    d.   installation and maintenance of a standpipe system.

    e.   building compartmentation.

70. **In regard to stair shafts in high-rise buildings:**

    a.   Scissors-type stairs consist of two sets of stairs in a common stair shaft.

    b.   Scissors-type stairs may alternate floors with each set of stairs in the stair shaft.

    c.   Stairs are located to enable total evacuation of a high-rise building.

    d.   Only a and b.

    e.   All of the above.

71. **High-rise buildings constructed today use computers that**

    a.   can assist the fire department operation by monitoring the protective systems in the building.

    b.   are flawless in defining building problems compared to occupants.

    c.   can provide a printout, or history, of what has occurred in the building.

    d.   only a and c.

    e.   all of the above.

72. **Two basic design concepts for horizontal floor separations in high-rise buildings are**

    a.   open and shut areas.

    b.   core and noncore areas.

    c.   compartmentation and open area.

    d.   only b and c.

    e.   all of the above.

73. **In regard to elevator usage by firefighters in high-rise buildings under fire conditions:**

    a.   Portable radio transmissions can affect electronic controls on some elevators.

    b.   If any doubt about the safe use of the elevator exists, climb the stairs.

    c.   Elevators always should be used so rescues can be performed in a timely manner.

    d.   Only a and b.

    e.   All of the above.

74. **Reflex or lag time refers to the time it takes**

    a.   to respond to the incident scene.

    b.   between receiving an order and accomplishing it.

    c.   for the Incident Commander to give orders to units on the scene.

    d.   only a and b.

    e.   all of the above.

75. **When using hose-lines at a high-rise fire**

    a.   firefighters may experience difficulty in advancing hose-lines down hallways.

    b.   access stairs may be used to stretch hose-lines to the floor above.

    c.   there is no need to stretch hose-lines to check areas adjacent to the fire area.

    d.   only a and b.

    e.   all of the above.

76.   **Poke-throughs in high-rise buildings**
- a.   are holes created between floors for utilities to pass through.
- b.   must be properly fire-stopped.
- c.   are not a problem for firefighters.
- d.   only a and b.
- e.   all of the above.

77.   **The best method for determining wind direction at a high-rise fire is**
- a.   checking with the National Weather Service.
- b.   breaking out a window in a fire area.
- c.   breaking out a window on the floor below the fire area.
- d.   only a and b.
- e.   all of the above.

78.   **Rapid intervention crews at a high-rise fire report to the**
- a.   Safety Officer.
- b.   Operations Section Chief.
- c.   Incident Commander.
- d.   only a and c.
- e.   all of the above.

79.   **At a high-rise fire, base is established**
- a.   on the interior two floors below staging.
- b.   on the exterior, a minimum of 100 feet from the fire building.
- c.   on the interior, adjacent to the lobby.
- d.   on the exterior, a minimum of 200 feet from the fire building.
- e.   none of the above.

80.   **When confronted with hazardous materials, a fixed facility can offer some benefits to initial responders. There may be fire protective systems in place to assist in mitigation. This could include**
- a.   sprinklers.
- b.   special extinguishing systems.
- c.   private hydrants.
- d.   ventilation systems.
- e.   all of the above.

81.   **At a hazardous materials incident setting up hot, warm, and cold zones can assist firefighter safety. Size consideration for the zones should be based upon**
- a.   wind direction.
- b.   terrain.
- c.   accessibility.
- d.   only a and b.
- e.   all of the above.

82. **At a hazardous materials incident setting up hot, warm, and cold zones can assist firefighter safety. Size consideration for the zones should be based upon**
    a. vapor clouds that exist.
    b. any explosion potential.
    c. the response direction of the fire department.
    d. only a and b.
    e. all of the above.

83. **The Incident Commander at a hazardous materials incident must**
    a. set up a command post in the cold zone.
    b. ensure that initial responding units operate within the realm of their training and capabilities.
    c. coordinate with police to establish site control.
    d. only a and b.
    e. all of the above.

84. 1) **Evacuation means removing everyone from the area of concern to a safe location. This method can be employed if it can be controlled and there is sufficient time and resources.**

    2) **In-place protection means keeping people indoors in the affected area and having them remain there until the danger passes.**
    a. Both statements are true.
    b. Both statements are false.
    c. Only statement number one is true.
    d. Only statement number two is true.

85. **At refineries and tank farms the covered floating roof tank**
    a. has a floating roof and a permanent solid cone roof.
    b. is permanently attached to the sidewalls with a weak shell joint.
    c. can be distinguished by vents near the top of the tank on the side walls.
    d. are usually free from ignitable vapors.
    e. all of the above.

86. **A terrorist incident includes the fact that**
    a. it will almost always involve a criminal activity.
    b. the amount of actual damage is a secondary concern of the terrorists.
    c. the primary concern of the terrorists is the psychological impact of the attack.
    d. only a and c.
    e. all of the above.

87. **Since the role of the emergency operator and dispatcher are so significant for early recognition and identification of a suspected terrorist event**

   a. dispatch centers must develop a list of questions to assist them in identification of a terrorist event.
   b. training for dispatchers should include signs and symptoms that could indicate the possibility of terrorism.
   c. dispatch centers should develop a list of locations or various types of occupancies that could trigger a warning.
   d. only a and c.
   e. all of the above.

88. **The earlier the recognition of the possibility of a terrorist event by firefighters, the faster safeguards can be initiated. These safeguards should include**

   a. approaching the suspected area from uphill and upwind.
   b. donning all protective equipment.
   c. attempting to cover all exposed skin.
   d. only b and c.
   e. all of the above.

89. **Depending upon the type of attack by terrorists there is the potential for a mass casualty incident due to**

   a. massive injuries due to a bomb.
   b. complex medical injuries due to a chemical attack.
   c. the detonation of a secondary device.
   d. only a and c.
   e. all of the above.

90. **Firefighters when operating at a suspected terrorist incident in the hot zone must wear their self-contained breathing apparatus. The wearer**

   a. may remove his or her mask upon entering the warm zone.
   b. may remove his or her mask upon entering the cold zone.
   c. must exit the warm zone and be decontaminated before removing his or her mask.
   d. must exit the hot zone and be decontaminated before removing his or her mask.
   e. none of the above.

91. **A return to normal is the last phase of a terrorist event. What occurs during the return to normal is**

   a. restoration of equipment.
   b. securing replacements for equipment that has been contaminated.
   c. critical incident stress debriefing.
   d. only a and b.
   e. all of the above.

92. **In regards to a dirty bomb:**
   a. A dirty bomb is a "homemade bomb."
   b. Uses conventional explosives and contains radioactive material.
   c. The bomb's purpose is to frighten people.
   d. Cell phones can be used to trigger the bomb.
   e. All of the above.

93. **The first arriving fire department chief officer at a clandestine drug lab will confer with the Police Incident Commander and:**
   a. Assist in determining the extent of the area to be evacuated.
   b. Advise on the size of the boundaries between hot, warm, and cold zones.
   c. Ensure that a sufficient number of hose-lines are stretched to unmanned monitors
   d. Only b and c
   e. All of the above

94. **Tornadoes are based on the Enhanced Fujita (EF) scale in the United States. How many categories are there?**
   a. 3.
   b. 4.
   c. 5.
   d. 6.
   e. 7.

95. **Assigning a new recruit to an experienced firefighter is part of the recruit's development process. Critiquing what occurred should take place after each incident. The explanations will assist the recruit in**
   a. understanding the importance of assignments.
   b. learning the teamwork needed for safe and efficient operations.
   c. his or her professional development.
   d. correcting mistakes and developing good fireground habits.
   e. all of the above.

96. **The final report on a formal critique would contain what size-up information?**
   a. Dispatch information received.
   b. Conditions observed during the fire.
   c. Problems encountered.
   d. Only b and c.
   e. All of the above.

97. **The final report on a formal critique would contain what safety information:**
   a. What were the safety issues?
   b. Was a Safety Officer assigned?
   c. What problems confronted the Safety Officer?
   d. Only b and c.
   e. All of the above.

98.   **The final component of the report at a formal critique should be**

     a.   discipline.
     b.   lessons learned.
     c.   finding out who made the most mistakes.
     d.   only a and b.
     e.   all of the above.

99.   **When dealing with serious civilian injuries firefighters must**

     a.   block out the emotional aspects of mutilation and human suffering.
     b.   continue to extricate and treat the injured or remove the dead.
     c.   often operate under circumstances that would be overwhelming to others.
     d.   only a and b.
     e.   all of the above.

100.   **Ground rules should be set whereby critical incident stress debriefing is mandatory for certain situations. Mandatory critical incident stress debriefing should include**

     a.   the death or serious injury to a firefighter.
     b.   the death of a child.
     c.   mass casualty incidents.
     d.   only a and b.
     e.   all of the above.

ANSWER KEY Study Guide 8 for Chapters 6 through 11

Question	Answer	Page Reference	Question	Answer	Page Reference
1	B	210	26	E	266
2	A	210	27	E	269
3	D	211	28	E	269
4	B	217	29	D	270 – 271
5	B	219	30	E	271
6	C	220	31	E	275
7	C	222	32	D	277
8	E	223	33	E	278 – 279
9	E	225	34	C	280
10	E	229	35	E	282
11	E	229 – 230	36	D	284
12	E	230 – 232	37	D	300
13	E	231 – 232	38	D	300
14	D	233	39	B	320
15	E	235	40	E	336
16	B	240	41	D	346
17	E	242	42	D	346
18	E	243	43	B	347
19	E	248	44	C	350
20	E	250	45	B	350
21	C	250	46	E	351 – 352
22	D	251	47	A	352
23	C	252	48	D	352
24	D	263	49	E	367 – 368
25	E	264	50	D	369

ANSWER KEY Study Guide 8 for Chapters 6 through 11

Question	Answer	Page Reference	Question	Answer	Page Reference
51	E	374	76	D	441
52	E	385	77	C	442
53	D	400	78	B	447
54	D	403	79	D	449
55	A	404	80	E	464
56	E	404	81	E	466
57	D	405	82	D	466
58	E	406	83	E	468 - 469
59	A	405 - 406	84	A	469
60	D	406 - 407	85	E	474
61	D	409	86	E	485
62	C	411	87	E	486
63	B	412	88	E	486
64	E	418	89	E	487
65	E	423	90	D	490
66	E	424	91	E	491
67	E	424 – 425	92	E	495
68	B	430	93	E	505
69	C	430	94	D	515
70	D	432 - 433	95	E	529
71	D	435	96	E	530
72	C	435	97	E	531
73	D	437	98	B	531
74	B	438	99	E	532
75	D	439 – 440	100	E	533

Study Guide 9 Test for all Chapters
F I N A L

1.   **Fires involving combustible metals: aluminum, magnesium, titanium, sodium, and potassium would be classified as:**
   a.   Class A fires.
   b.   Class B fires.
   c.   Class C fires.
   d.   Class D fires.
   e.   Class E fires.

2.   **Routine training with mutual aid departments enables members to**
   a.   learn new radio procedures.
   b.   bond friendships and share firefighting experiences.
   c.   discover other fire department's rank structure.
   d.   learn little that is beneficial to the individual's fire department.
   e.   none of the above.

3.   **Preincident plans should be prepared for target hazards that include buildings or processes for**
   a.   situations that could create a conflagration hazard.
   b.   facilities or structures that have a high frequency of fires.
   c.   facilities whose loss would have a large economic impact on the community.
   d.   none of the above.
   e.   all of the above.

4.   1)   **An index card system used for preplans can be difficult to access at an incident scene.**
      2)   **A booklet containing comprehensive data can back up an index card system that is used for preplans.**
   a.   Both statements are true.
   b.   Both statements are false.
   c.   Only statement number one is true.
   d.   Only statement number two is true.

5.   1)   **If the National Fire Academy's fire flow requirements for water supply exceed the fire flow capability of available resources, a defensive mode of operation usually is required.**

        2)   **Situations will occur where fire is attacking lightweight structural components and, though there is a sufficient water supply and resources at the scene, the conditions will be too dangerous for an offensive attack.**

        a.   Both statements are true.
        b.   Both statements are false.
        c.   Only statement number one is true.
        d.   Only statement number two is true.

6.   1)   **Command must be able to predict changes in the incident scene while evaluating the effectiveness of the firefighting efforts.**

        2)   **Firefighter safety demands that they be an integral part of a "known" plan to prevent firefighters from being placed in dangerous positions.**

        a.   Both statements are true.
        b.   Both statements are false.
        c.   Only statement number one is true.
        d.   Only statement number two is true.

7.   1.   **Constant re-evaluation of the incident is necessary to ensure that the Incident Commander's goals are being accomplishing.**

        2.   **The Incident Commander at most incidents will oversee the tactical operations.**

        a.   Both statements are true.
        b.   Both statements are false.
        c.   Only statement number one is true.
        d.   Only statement number two is true.

8.   1)   **The Incident Commander at a major incident must make use of his or her staff by delegating tactical decisions to subcommands.**

        2)   **The Incident Commander should not give just an overall objective to subcommands but specific, point-by-point orders to accomplish the task.**

        a.   Both statements are true.
        b.   Both statements are false.
        c.   Only statement number one is true.
        d.   Only statement number two is true.

9. 1) **The rotation of companies from staging to relieve units operating at the scene enables more firefighters to gain experience.**
    2) **Staging can be used in a variety of ways and various levels.**
    a. Both statements are true.
    b. Both statements are false.
    c. Only statement number one is true.
    d. Only statement number two is true.

10. **The Logistics Section Chief has responsibility for the:**
    a. Communications Unit.
    b. Medical Unit.
    c. Service and Support Branches.
    d. Ground Support Unit.
    e. All of the above.

11. **To accomplish Mobile Command the first arriving officer**
    a. gives an initial status report.
    b. gives orders for the incoming units.
    c. identifies and assumes Command.
    d. all of the above.
    e. none of the above.

12. 1) **When giving dimensions of a structure for an incident status report always give the depth first, then the width of a building.**
    2) **A comprehensive initial status report allows the chief to envision the magnitude of the problem.**
    a. Both statements are true.
    b. Both statements are false.
    c. Only statement number one is true.
    d. Only statement number two is true.

13. **The length of time for an operational period is determined by:**
    a. Incident Commander.
    b. Operations Section Chief.
    c. Planning Section Chief.
    d. Logistics Section Chief.
    e. All of the above.

14. **Cue-based decision making**
    a. is readily recallable information.
    b. ties together past and present events.
    c. is accomplished by unified discussions at the incident.
    d. only a and b.
    e. all of the above.

15.   **1)**   **The best method of remaining proactive at an incident scene is through the use of a logical thought process.**
      2)   **The determination of incident priorities, in conjunction with size-up, assists in the development of strategy and tactics.**
      a.   Both statements are true.
      b.   Both statements are false.
      c.   Only statement number one is true.
      d.   Only statement number two is true.

16.   **1)**   **Size-up is an evaluation process that reviews all critical factors that could have a positive or negative impact on an incident.**
      2)   **Size-up starts in the preplanning stages.**
      a.   Both statements are true.
      b.   Both statements are false.
      c.   Only statement number one is true.
      d.   Only statement number two is true.

17.   **Assuming that sufficient resources are available, what will dictate whether a fire can be controlled?**
      a.   Pump capacity.
      b.   Water supply.
      c.   Ventilation.
      d.   The type of fire department.
      e.   The classification type of the fire.

18.   **Information that should be available at an incident scene regarding water supply includes**
      a.   a system to quickly determine hydrant locations.
      b.   the size of the water mains.
      c.   location of drafting sites.
      d.   only a and b.
      e.   all of the above.

19.   **Knowledge of the location of fire walls, fire doors and whether they are kept closed will be useful**
      a.   in predicting fire spread.
      b.   when deciding where an arsonist may set a fire.
      c.   in the placement of units to contain a fire.
      d.   only a and c.
      e.   all of the above.

20. 1) **The exposure problem addresses two basic areas, internal and external.**

   2) **With internal exposures, contents in the immediate and adjoining areas must be considered. When dealing with exterior exposures, the direction in which a fire may spread and to which other structures must be considered.**

   a. Both statements are true.
   b. Both statements are false.
   c. Only statement number one is true.
   d. Only statement number two is true.

21. **Firefighters operating at residential structures at night**

   a. will encounter more locked doors.
   b. must perform more forcible entry.
   c. will find restricted vision due to darkness.
   d. only a and c.
   e. all of the above.

22. **The systematic deployment of strategy should be considered a tool since**

   a. knowledge is needed to evaluate the information gathered.
   b. experience lets us draw upon actions that have been successful in the past.
   c. training allows us to be proficient in the performance of our duties.
   d. training eliminates unnecessary actions.
   e. all of the above

23. **A critical point to remember is that an initial responder will basically be confronted with only four of the seven strategic considerations. Those initial concerns are:**

   a. Rescue, ventilation, confinement, and overhaul.
   b. Rescue, confinement, extinguishment and overhaul.
   c. Rescue, confinement, ventilation, and extinguishment.
   d. Rescue, exposures, confinement and ventilation.
   e. Rescue, exposures, confinement, and extinguishment.

24. **Extinguishment of a fire involves**

   a. the knocking down of all visible fire.
   b. exposing hidden fire during the overhaul stage.
   c. removing undamaged property to a safe location.
   d. only a and b.
   e. all of the above.

25. 1) **The attacking of a fire in a single room or small area often accomplishes both confinement and extinguishment.**

   2) **Extinguishment calls for the judicious use of water.**

   a. Both statements are true.
   b. Both statements are false.
   c. Only statement number one is true.
   d. Only statement number two is true.

26.   **The 1½-inch or 1¾-inch hose-line is**
- a.   the workhorse of the fire service.
- b.   easy to maneuver in confined spaces.
- c.   always effective on fighting fires in commercial properties.
- d.   only a and b.
- e.   all of the above.

27.   **If a hose-line cannot advance it may be due to**
- a.   too many firefighters spaced along the hose-line.
- b.   inadequate ventilation.
- c.   the hose-line being too small.
- d.   only b and c.
- e.   all of the above.

28.   **Building standpipe systems that provide 2½-inch hose-line connections and are intended for use by firefighters for full-scale firefighting are classified as:**
- a.   Class A.
- b.   Class B.
- c.   Class C.
- d.   Class 1.
- e.   Class 2.

29.   **A primary search of a fire building involves**
- a.   a coordinated and systematic search via the interior.
- b.   checking first the area around the fire area and directly above the fire.
- c.   removal of the most seriously endangered occupants first.
- d.   only a and c.
- e.   all of the above.

30.   **In regard to search and rescue:**
- a.   A hose-line is never used in search and rescue.
- b.   Members always must use a hose-line when performing search and rescue.
- c.   The optimum search is made under the protection of a hose-line.
- d.   Only b and c.
- e.   All of the above.

31.   **The initial Incident Commander arriving on a fire scene where many people are endangered must decide**
- a.   where the fire department's efforts will do the most good.
- b.   what additional resources are responding.
- c.   whether a hose-line must be stretched to contain a fire to a specific area.
- d.   only a and b.
- e.   all of the above.

32.   1)   **When a ladder is placed for rescue from a window, the top of the ladder must protrude into the window opening to assist in mounting the ladder from the interior**

      2)   **A ladder placed for a firefighter to enter a window can be alongside the window and about three rungs above the windowsill**

      a.   Both statements are true.
      b.   Both statements are false.
      c.   Only statement number one is true.
      d.   Only statement number two is true.

33.   **Firefighters involved in roof operations should ensure that there are at least two independent means of egress from the roof. These include**

      a.   two ladders raised by the fire department.
      b.   a combination of a raised ladder and a fire escape or a fire tower.
      c.   a raised ladder and the roof of an adjoining building.
      d.   only a and b.
      e.   all of the above.

34.   1)   **Ventilation can be as basic as firefighters opening doors and windows to allow free circulation of outside air.**

      2)   **Properly performed ventilation enables firefighters to extinguish a fire quickly and efficiently.**

      a.   Both statements are true.
      b.   Both statements are false.
      c.   Only statement number one is true.
      d.   Only statement number two is true.

35.   **In regards to rooftop solar electrical panels firefighters should know**

      a.   they are a source of electrical energy.
      b.   moonlight cannot energize these panels.
      c.   there may be a battery storage system.
      d.   they may impede roof top ventilation.
      e.   all of the above.

36.   1)   **Combustible dust lying on a heated surface is subject to ignition due to carbonization of the dust.**

      2)   **Dust explosions usually occur in pairs. The initial explosion may not cause substantial damage, but the secondary explosion is usually devastating.**

      a.   Both statements are true.
      b.   Both statements are false.
      c.   Only statement number one is true.
      d.   Only statement number two is true.

37. **The most common method(s) of protecting steel from the heat of a fire is/are**
   a. membrane protection.
   b. sprayed on protection.
   c. encasement.
   d. only b and c.
   e. all of the above.

38. 1) **The strength of concrete depends upon how it is supported. Reinforced concrete is very strong if properly constructed, and if the steel reinforcement is well protected.**
   2) **Concrete does not absorb the heat of a fire readily, nor retain that heat.**
   a. Both statements are true.
   b. Both statements are false.
   c. Only statement number one is true.
   d. Only statement number two is true.

39. **Ordinary constructed buildings contain exterior**
   a. wooden walls and interior floors and roofs constructed of wood.
   b. wooden walls and interior floors and roofs constructed of steel.
   c. masonry walls and interior floors and roofs constructed of wood.
   d. masonry walls and interior floors and roofs constructed of steel.
   e. none of the above.

40. **To support the weight of the masonry wall above an opening made for windows and doors requires that the load be transferred to the sides of the opening. This is accomplished by installing**
   a. a lintel.
   b. an arch.
   c. a door or window frame.
   d. only a and b.
   e. all of the above.

41. **A fire wall in heavy timber construction is customarily**
   a. a bearing masonry wall.
   b. a barrier to fire.
   c. any wall separating two distinct fire areas.
   d. only a and b.
   e. all of the above.

42. **Modifications made to heavy timber buildings include makeshift offices that can create problems for firefighters since**
   a. they may be built beneath the sprinkler system and not contain sprinklers in these rooms.
   b. portable heaters used in these makeshift offices can overload the electrical wiring.
   c. an accumulation of paperwork and improper storage may set the stage for a fire to be well involved before activation of the building's alarm system.
   d. only a and b.
   e. all of the above.

43. **When fighting fires in heavy timber buildings a decision on which direction to fight the fire must consider**
   a. access.
   b. fire wall locations.
   c. minimizing building loss.
   d. only a and b.
   e. all of the above.

44. **The use of a blitz attack when fighting a fire in heavy timber constructed buildings**
   a. can knock down large amounts of fire while interior hose-lines are being stretched.
   b. is of little value due to the size of these buildings.
   c. should be used on every fire in these buildings.
   d. only a and b.
   e. none of the above.

45. **The characteristics of post and beam frame construction are**
   a. the framing system uses posts as the vertical members.
   b. the framing system uses beams as the horizontal members.
   c. the framing system is connected by rigid joints to support the structure.
   d. only a and b.
   e. all of the above.

46. **Nominal sized lumber**
   a. is smaller than the named sized piece of material.
   b. contains less support than wood of full dimensional size.
   c. is stronger than wood of full dimensional size.
   d. only a and b.
   e. all of the above.

47. **Pressure treated or "green lumber" is wood permeated with various chemicals to resist attack from moisture and insects. The common treatments use**

a. phenol.
b. arsenic.
c. tars.
d. only a and b.
e. all of the above.

48. **Brick veneer is a single course or wythe of brick attached to a building**

a. for aesthetic purposes.
b. to support the wall it is attached to.
c. to support floor and roof joist.
d. only b and c.
e. all of the above.

49. **The split ring connector used in heavy timber truss is designed to**

a. spread the pressure over a wide area rather than strictly on the bolt.
b. split the top and bottom chord.
c. relieve some of the shear placed on the bolt.
d. only a and c.
e. all of the above.

50. **A major factor associated with using steel in any truss is**

a. it withstands the heat of a fire for a prolonged period of time.
b. it fails readily under fire conditions.
c. it does not conduct heat.
d. only a and c.
e. all of the above.

51. **The only rule of thumb that is accurate in predicting wall collapse that will ensure firefighter safety is**

a. a wall will fall one-third of its height.
b. a wall can fall 100 percent of its height.
c. a solid wall will crumble as it collapses and will fall only 50 percent its height.
d. only a and c.
e. none of the above.

52. **The amount of fire and where it is burning can be used as a collapse indicator. Continued or heavy fire for 15 to 20 minutes is a reference point to indicate collapse potential. This timeframe considers**

a. that the fire is attacking solid wood structural members.
b. buildings using lightweight components.
c. that the fire should have out within this time-frame.
d. only a and c.
e. all of the above.

53. 1) **Fire burning in an area containing lightweight wood truss will directly attack these structural members and they will not withstand direct flame impingement.**
     2) **Steel components and sheet metal surface fasteners will resist the heat of a fire and remain stable and not collapse under intense fire conditions.**
     a. Both statements are true.
     b. Both statements are false.
     c. Only statement number one is true.
     d. Only statement number two is true.

54. **Sustained heat can cause visible spalling of a brick or concrete wall. When spalling occurs**
     a. the wall is weakened.
     b. it has little impact on the wall, since it is of masonry construction.
     c. structural steel or steel rods in concrete can be exposed to the heat of the fire.
     d. only a and c.
     e. all of the above.

55. **Staircases that are out of level can indicate that a building has shifted. If the staircase shifts enough it may not support the weight of a firefighter. Firefighters finding this condition should**
     a. keep close to the wall when ascending or descending the stairs.
     b. place a portable ladder up the staircase.
     c. immediately call for power saws to cut out the staircases before they collapse.
     d. only a and b.
     e. all of the above.

56. 1) **The live load in a building can constantly change.**
     2) **The dead load is the weight of the building components that includes the masonry, lumber, light fixtures, piping, molding, windows, etc.**
     a. Both statements are true.
     b. Both statements are false.
     c. Only statement number one is true.
     d. Only statement number two is true.

57. **When confronted with a collapsed building, and a need to locate victims, the Incident Commander should ensure**
     a. the verification of any sounds being made by the victims to get as exact a location as possible.
     b. that contact is maintained between victims and searchers.
     c. the immediate removal of victims, and not be too concerned about their condition.
     d. only a and b.
     e. all of the above.

58. The Safety Officer's authority includes

a. being able to bypass the chain of command when he or she wants to.

b. being able to bypass the chain of command when necessary to remove firefighters from imminent danger.

c. being able to bypass the chain of command when he or she outranks the officer assigned to a division or group.

d. only b and c.

e. all of the above.

59. 1) **Once a hazard is identified, it can be controlled or mitigated.**

2) **Risks are the chances people take in relationship to hazards.**

a. Both statements are true.

b. Both statements are false.

c. Only statement number one is true.

d. Only statement number two is true.

60. 1) **A cellar can be one large open area, which can be more conducive to firefighting.**

2) **When attacking a cellar fire the reach of a stream will be reduced if the height of storage reaches near the ceiling.**

a. Both statements are true.

b. Both statements are false.

c. Only statement number one is true.

d. Only statement number two is true.

61. A cellar fire may be recognized by

a. heavy smoke coming only from the upper floors of a building.

b. the presence of heat and smoke at the first-floor level and the absence of visible fire.

c. smoke emitting from the baseboards on lower floors and banking down on the top floor.

d. only b and c.

e. all of the above.

62. The presence of what appears to be light smoke in a basement or cellar

a. will indicate a situation of little concern.

b. allows firefighters to immediately remove their self-contained breathing apparatus.

c. still requires the continued use of self-contained breathing apparatus.

d. only a and b.

e. all of the above.

63. **Water use and its subsequent build-up in a basement or cellar**

    a.   can prevent firefighters from finding their hose-line to use it as a guideline to exit a fire area.

    b.   can spread storage about, burying hose-line beneath it.

    c.   will not be a problem, since all fire departments use good water management at cellar fires.

    d.   only a and b.

    e.   all of the above.

64. **In regard to ventilation at a cellar fire**

    a.   deadlights in the sidewalk can be broken out to effect ventilation.

    b.   breaking out the material under showcase windows in stores will assist in ventilation of a cellar.

    c.   ventilation in fighting a cellar fire can be effected by opening upper floor windows.

    d.   only a and b.

    e.   all of the above.

65. **The term "garden apartments" infers**

    a.   that there are gardens surrounding the apartment buildings.

    b.   that the building is set back from the roadway.

    c.   that these types of buildings were originally found along the beer gardens in Germany.

    d.   only a and b.

    e.   all of the above.

66. **Characteristics of garden apartments include**

    a.   four apartments per floor can be found in a section.

    b.   fire resistive construction is usually employed.

    c.   sections are usually constructed adjacent to other sections.

    d.   only a and c.

    e.   all of the above.

67. **A roof overhanging the sidewalls of a building creates**

    a.   eaves.

    b.   bearing walls.

    c.   nonbearing walls.

    d.   only a and b.

    e.   all of the above.

68. **A poke-through**

    a.   is an opening in draft-stopping material.

    b.   negates compartmentation.

    c.   can create a large non-fire-stopped area within a concealed space.

    d.   only a and c.

    e.   all of the above.

69.   **In regard to windows in garden apartments**

    a.   windows may be sliding glass windows.

    b.   windows may be placed high in the wall hindering rescues.

    c.   windows are always large and allow easy egress under emergency conditions.

    d.   only a and b.

    e.   all of the above.

70.   **1)**   **When fighting fires in garden apartments the Incident Commander can gain information from occupants and by observing conditions on the location and the approximate extent of the fire.**

       2)   **When fighting fires in garden apartments the Incident Commander should monitor smoke conditions to see the amount and density of the smoke in the various apartments in the section.**

    a.   Both statements are true.

    b.   Both statements are false.

    c.   Only statement number one is true.

    d.   Only statement number two is true.

71.   **Characteristics of row houses and town houses in general include that**

    a.   they contain a common party wall.

    b.   the front and rear walls are normally nonbearing walls.

    c.   the party walls are normally nonbearing walls.

    d.   only a and b.

    e.   all of the above.

72.   **Concerning row houses in general**

    a.   they are separated into many small rooms.

    b.   the problems increase as the structures get bigger.

    c.   interior spread of fire can be contained easily.

    d.   only a and b.

    e.   all of the above.

73.   **Row house construction often allows joists laid in a common party wall to abut the joists in the adjoining buildings. Firefighters should know that**

    a.   this is never a concern unless heavy smoke is showing in the adjoining buildings.

    b.   it can permit conduction of fire from one building to the next.

    c.   it may necessitate opening walls and ceilings in exposed buildings.

    d.   only b and c.

    e.   all of the above.

74. **Problems associated with renovated buildings include**
a. sprinkler systems that are out of service and can allow a fire to get past the incipient stage.
b. firestops may be removed.
c. problems no different than any other building.
d. only a and b.
e. all of the above.

75. **In regard to vacant building fires**
a. fireground problems will be accentuated in these properties.
b. they cause more firefighter injuries and deaths than other structural fires.
c. they are easier types of fires to attack, since there is never any life hazard.
d. only a and b.
e. all of the above.

76. **1)** **Vacant building fires must be handled in the safest manner possible. It is not realistic to fight every vacant building fire in a defensive mode.**

2) **An exterior attack on fires past the initial stages can be a successful tactic. The exterior knockdown or blitz attack with master streams can be followed by an interior hose-line team, if deemed safe by the Safety Officer.**
a. Both statements are true.
b. Both statements are false.
c. Only statement number one is true.
d. Only statement number two is true.

77. **The wildland acronym LCES stand for:**
a. Listen, Call, Enter, and Safety zones
b. Look, Call, Enter, Safe exits
c. Lookouts, Communications, Escape routes, Safety zones
d. Let's Confirm Everyone's Safety
e. None of the above

78. **The leading contributing factor of major fires in churches is**
a. the large area on the interior of the building.
b. delayed detection or delayed notification of the fire department.
c. poor firefighting techniques.
d. only a and b.
e. all of the above.

79. **In regards to school violence involving an active shooter, the fire department will generally:**
   a. Assist law enforcement in subduing the active shooter.
   b. Allow fire department apparatus to be used as shields for police to approach the incident.
   c. Operate in secured areas only.
   d. Allow police officers to utilize fire department clothing to act as a decoy.
   e. All of the above.

80. **In regard to the operation of a sprinkler system when fighting fires in commercial buildings and warehouses:**
   a. Firefighters should immediately shut down the sprinkler system to reduce water damage.
   b. The firefighter shutting down the sprinkler system must remain at the valve location.
   c. The firefighter shutting down the sprinkler system does not need to remain at the valve location, but must maintain communications should the system need to be turned on.
   d. Only a and b.
   e. All of the above.

81. **In regard to commercial buildings and warehouses that contain masonry fire walls:**
   a. These fire walls are customarily self-supporting.
   b. Horizontal openings can be protected by fire doors.
   c. Fire wall is another term for nonbearing wall.
   d. Only a and b.
   e. All of the above.

82. **In regard to a strip mall:**
   a. The underside of the roof may be painted and left exposed to the interior of the store.
   b. Partition walls may end on the underside of the ceiling.
   c. A fire that extends above the suspended ceiling can spread to adjacent stores and attack the entire roof area.
   d. Only a and b.
   e. All of the above.

83. **Downtown enclosed malls often have trash chutes. These trash chutes**
   a. can have fires occurring within them.
   b. can spread smoke throughout the mall area when involved in fires.
   c. are really not a problem since the sprinkler system will immediately extinguish any fire.
   d. only a and b.
   e. all of the above.

84.   **Ventilation considerations at supermarket fires can include:**

a.   Utilizing skylights or operable roof hatches.
b.   Removal of trailers from loading platforms.
c.   Never breaking out the large front plate glass windows.
d.   Only a and b.
e.   All of the above.

85.   **Fires in modern high-rise buildings**

a.   generate much higher temperatures.
b.   are hotter, since the building materials do not absorb the heat of the fire.
c.   burn the same as those built in the early part of the 20th century.
d.   only a and b.
e.   all of the above.

86.   **There are various methods of marking hazardous materials. In the United States the regulation of labels and placards that must be attached to vehicles during transportation is the responsibility of?**

a.   Interstate Transportation Association.
b.   U.S. Department of Transportation.
c.   United Nations.
d.   Only b and c.
e.   All of the above.

87.   **1)**   **Many responses to hazardous materials incidents occur in laboratories within hospitals, research facilities, or industrial plants. These facilities have recirculation of air within the building that can rapidly spread the problem to other areas.**

   2)   **Fixed facilities dealing with hazardous materials can handle most problems that occur. They will request assistance when someone is injured or a spill, leak, or fire is out of control.**

a.   Both statements are true.
b.   Both statements are false.
c.   Only statement number one is true.
d.   Only statement number two is true.

88.   **On suspected hazardous materials responses**

a.   dispatch should transmit wind direction and speed.
b.   terrain need not be considered when approaching a suspected hazardous materials site.
c.   positioning of apparatus must consider the possibility of contamination.
d.   only a and c.
e.   all of the above.

89. 1) **A hazardous materials incident requires a more cautious and deliberate size-up than most other types of responses. One consideration is "what if the fire department does nothing."**

    2) **The data in the *Emergency Response Guidebook* can be used by the first-arriving fire department units for setting up the initial zones and for the evacuation of civilians.**

    a. Both statements are true.
    b. Both statements are false.
    c. Only statement number one is true.
    d. Only statement number two is true.

90. **The term BLEVE refers to**
    a. Big Levitation Expected Very Early.
    b. Boiling Liquid Explosion of Vapor Exponents.
    c. Boiling Liquid Expected Vapor Explosion.
    d. Boiling Liquid Expanding Vapor Explosion.
    e. None of the above.

91. **To adequately recoup funds and equipment that are expended at a hazardous materials incident, fire departments must**
    a. buy from vendors approved by the Federal government.
    b. use all of the equipment for which they seek reimbursement.
    c. be meticulous in their recordkeeping.
    d. only a and b.
    e. all of the above.

92. 1) **Evacuation means removing everyone from the area of concern. This method can be employed if it can be controlled and if there is sufficient time and resources.**

    2) **In-place protection means moving or keeping people indoors in the affected area until the danger passes.**

    a. Both statements are true.
    b. Both statements are false.
    c. Only statement number one is true.
    d. Only statement number two is true.

93. **The Federal Bureau of Investigation recognizes two categories of terrorism:**
    a. National and international.
    b. Home country and foreign country.
    c. Domestic and international.
    d. Only a and b.
    e. All of the above.

94. 1)   **The United States Department of Justice describes terrorism in part as "a violent act or an act dangerous to human life, in violation of criminal laws of the United States."**

     2)   **Domestic terrorism is classified as those groups or individuals whose terrorist acts are directed at elements of our government or population with foreign direction.**

     a.   Both statements are true.
     b.   Both statements are false.
     c.   Only statement number one is true.
     d.   Only statement number two is true.

95. **The weapon of choice of the greatest number of terrorist acts involves**

     a.   chemicals.
     b.   biological agents.
     c.   explosives.
     d.   only a and b.
     e.   all of the above.

96. **Some indicators of methamphetamine lab fires are:**

     a.   Unusual flame color due to the chemicals involved.
     b.   Violent reactions to water streams.
     c.   Strong chemical odors.
     d.   Empty chemical containers.
     e.   All of the above.

97. **At a natural disaster an immediate damage assessment will need to be conducted in the initial stages of the disaster. The information gathered can be used:**

     a.   As part of size-up and for initial reports.
     b.   To assist in establishing an incident management organization.
     c.   In developing incident objectives and strategy development.
     d.   To determine tactical operations.
     e.   All of the above.

98. **Commanding officers should place mistakes or errors into two categories**

     a.   good and bad.
     b.   commission and omission.
     c.   great and not so great.
     d.   okay and overlooked.
     e.   none of the above.

99. **A written report of a formal critique would be concerned with communications at an incident in the following areas:**

     a.   Were any problems found to arise?
     b.   Did dispatch receive proper and timely reports?
     c.   Did everyone have a radio?
     d.   Only a and b.
     e.   All of the above.

100. **1)** **Critical incident stress debriefing will reduce incident stress. Mental health professionals and peer counselors are needed for a formal debriefing.**
     2) **A formal critical incident stress debriefing can occur immediately after the incident, or whenever it is convenient, and should include members who responded to the incident that are available.**
     a.   Both statements are true.
     b.   Both statements are false.
     c.   Only statement number one is true.
     d.   Only statement number two is true.

ANSWER KEY Study Guide 9 - Final - All Chapters

Question	Answer	Page Reference	Question	Answer	Page Reference
1	D	4	26	D	135
2	B	11	27	D	136
3	E	12 – 13	28	D	139
4	D	13	29	E	142
5	A	19	30	C	143
6	A	56	31	E	145
7	A	57	32	D	147
8	C	59	33	E	148
9	A	62	34	A	152
10	E	69	35	E	156 – 157
11	D	74	36	A	165
12	D	76	37	E	173
13	A	84	38	C	173
14	D	90	39	C	178
15	A	91 – 92	40	D	182
16	A	93	41	D	183
17	B	99	42	E	184
18	E	99	43	E	185
19	D	101	44	A	185
20	A	102	45	E	187
21	E	105	46	D	188
22	E	106	47	D	189
23	D	107	48	A	189
24	D	108	49	D	190 – 191
25	A	108	50	B	192

ANSWER KEY Study Guide 9 - Final - All Chapters

Question	Answer	Page Reference	Question	Answer	Page Reference
51	B	211	76	A	300
52	A	216	77	C	312
53	C	217 – 218	78	B	348
54	D	220	79	C	386
55	D	222	80	B	398
56	A	226	81	D	399
57	D	232	82	E	404
58	B	240	83	D	410
59	A	242	84	D	419-420
60	A	263	85	D	430
61	D	264	86	B	464
62	C	265	87	D	465
63	D	265	88	D	465 - 466
64	D	266	89	A	466
65	B	268	90	D	467
66	D	268	91	C	469
67	A	269	92	A	469
68	E	269	93	C	484
69	D	269	94	C	484 – 485
70	A	270	95	C	485
71	D	275	96	E	508
72	D	277 – 278	97	E	518
73	D	279	98	B	529
74	D	283	99	D	530
75	D	297	100	C	533